Teacher Edition

Eureka Math
Grade 1
Module 5

Special thanks go to the Gordon A. Cain Center and to the Department of Mathematics at Louisiana State University for their support in the development of *Eureka Math*.

For a free *Eureka Math* Teacher Resource Pack, Parent Tip Sheets, and more please visit www.Eureka.tools

Published by the non-profit Great Minds

Copyright © 2015 Great Minds. No part of this work may be reproduced, sold, or commercialized, in whole or in part, without written permission from Great Minds. Non-commercial use is licensed pursuant to a Creative Commons Attribution-NonCommercial-ShareAlike 4.0 license; for more information, go to http://greatminds.net/maps/math/copyright. "Great Minds" and "Eureka Math" are registered trademarks of Great Minds.

Printed in the U.S.A.
This book may be purchased from the publisher at eureka-math.org

ISBN 978-1-63255-352-2

Eureka Math: A Story of Units **Contributors**

Katrina Abdussalaam, Curriculum Writer
Tiah Alphonso, Program Manager—Curriculum Production
Kelly Alsup, Lead Writer / Editor, Grade 4
Catriona Anderson, Program Manager—Implementation Support
Debbie Andorka-Aceves, Curriculum Writer
Eric Angel, Curriculum Writer
Leslie Arceneaux, Lead Writer / Editor, Grade 5
Kate McGill Austin, Lead Writer / Editor, Grades PreK–K
Adam Baker, Lead Writer / Editor, Grade 5
Scott Baldridge, Lead Mathematician and Lead Curriculum Writer
Beth Barnes, Curriculum Writer
Bonnie Bergstresser, Math Auditor
Bill Davidson, Fluency Specialist
Jill Diniz, Program Director
Nancy Diorio, Curriculum Writer
Nancy Doorey, Assessment Advisor
Lacy Endo-Peery, Lead Writer / Editor, Grades PreK–K
Ana Estela, Curriculum Writer
Lessa Faltermann, Math Auditor
Janice Fan, Curriculum Writer
Ellen Fort, Math Auditor
Peggy Golden, Curriculum Writer
Maria Gomes, Pre-Kindergarten Practitioner
Pam Goodner, Curriculum Writer
Greg Gorman, Curriculum Writer
Melanie Gutierrez, Curriculum Writer
Bob Hollister, Math Auditor
Kelley Isinger, Curriculum Writer
Nuhad Jamal, Curriculum Writer
Mary Jones, Lead Writer / Editor, Grade 4
Halle Kananak, Curriculum Writer
Susan Lee, Lead Writer / Editor, Grade 3
Jennifer Loftin, Program Manager—Professional Development
Soo Jin Lu, Curriculum Writer
Nell McAnelly, Project Director

Ben McCarty, Lead Mathematician / Editor, PreK–5
Stacie McClintock, Document Production Manager
Cristina Metcalf, Lead Writer / Editor, Grade 3
Susan Midlarsky, Curriculum Writer
Pat Mohr, Curriculum Writer
Sarah Oyler, Document Coordinator
Victoria Peacock, Curriculum Writer
Jenny Petrosino, Curriculum Writer
Terrie Poehl, Math Auditor
Robin Ramos, Lead Curriculum Writer / Editor, PreK–5
Kristen Riedel, Math Audit Team Lead
Cecilia Rudzitis, Curriculum Writer
Tricia Salerno, Curriculum Writer
Chris Sarlo, Curriculum Writer
Ann Rose Sentoro, Curriculum Writer
Colleen Sheeron, Lead Writer / Editor, Grade 2
Gail Smith, Curriculum Writer
Shelley Snow, Curriculum Writer
Robyn Sorenson, Math Auditor
Kelly Spinks, Curriculum Writer
Marianne Strayton, Lead Writer / Editor, Grade 1
Theresa Streeter, Math Auditor
Lily Talcott, Curriculum Writer
Kevin Tougher, Curriculum Writer
Saffron VanGalder, Lead Writer / Editor, Grade 3
Lisa Watts-Lawton, Lead Writer / Editor, Grade 2
Erin Wheeler, Curriculum Writer
MaryJo Wieland, Curriculum Writer
Allison Witcraft, Math Auditor
Jessa Woods, Curriculum Writer
Hae Jung Yang, Lead Writer / Editor, Grade 1

Board of Trustees

Lynne Munson, President and Executive Director of Great Minds
Nell McAnelly, Chairman, Co-Director Emeritus of the Gordon A. Cain Center for STEM Literacy at Louisiana State University
William Kelly, Treasurer, Co-Founder and CEO at ReelDx
Jason Griffiths, Secretary, Director of Programs at the National Academy of Advanced Teacher Education
Pascal Forgione, Former Executive Director of the Center on K-12 Assessment and Performance Management at ETS
Lorraine Griffith, Title I Reading Specialist at West Buncombe Elementary School in Asheville, North Carolina
Bill Honig, President of the Consortium on Reading Excellence (CORE)
Richard Kessler, Executive Dean of Mannes College the New School for Music
Chi Kim, Former Superintendent, Ross School District
Karen LeFever, Executive Vice President and Chief Development Officer at ChanceLight Behavioral Health and Education
Maria Neira, Former Vice President, New York State United Teachers

This page intentionally left blank

A STORY OF UNITS

GRADE

Mathematics Curriculum

GRADE 1 • MODULE 5

Table of Contents

GRADE 1 • MODULE 5

Identifying, Composing, and Partitioning Shapes

Module Overview 2

Topic A: Attributes of Shapes 9

Topic B: Part–Whole Relationships Within Composite Shapes 62

Topic C: Halves and Quarters of Rectangles and Circles 98

Topic D: Application of Halves to Tell Time 135

End-of-Module Assessment and Rubric 183

Answer Key 199

 Module 5: Identifying, Composing, and Partitioning Shapes

Grade 1 • Module 5
Identifying, Composing, and Partitioning Shapes

OVERVIEW

Throughout the year, students have explored part–whole relationships in many ways, such as their work with number bonds, tape diagrams, and the relationship between addition and subtraction. In Module 5, students consider part–whole relationships through a geometric lens.

In Topic A, students identify the defining parts, or attributes, of two- and three-dimensional shapes, building on their kindergarten experiences of sorting, analyzing, comparing, and creating various two- and three-dimensional shapes and objects (**1.G.1**). Using straws, students begin the exploration by creating and describing two-dimensional shapes without naming them. This encourages students to attend to and clarify a shape's defining attributes. In the following lessons, students name two- and three-dimensional shapes and find them in pictures and in their environment. New shape names are added to the students' repertoire, including *trapezoid, rhombus, cone,* and *rectangular prism*.

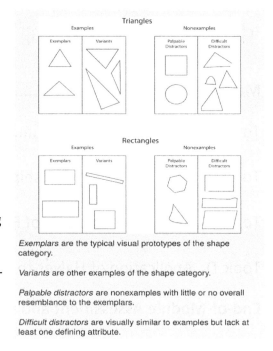

Exemplars are the typical visual prototypes of the shape category.

Variants are other examples of the shape category.

Palpable distractors are nonexamples with little or no overall resemblance to the exemplars.

Difficult distractors are visually similar to examples but lack at least one defining attribute.

In Topic B, students combine shapes to create a new whole: a composite shape (**1.G.2**). Students identify the name of the composite shape as well as the names of each shape that forms it. Students see that another shape can be added to a composite shape so that the composite shape becomes part of an even larger whole.

In Topic C, students relate geometric figures to equal parts and name the parts as halves and fourths (or quarters) (**1.G.3**). For example, students now see that a rectangle can be partitioned into two equal triangles (whole to part) and that the same triangles can be recomposed to form the original rectangle (part to whole). Students see that as they create more parts, decomposing the shares from halves to fourths, the parts get smaller.

The module closes with Topic D, in which students apply their understanding of halves (**1.G.3**) to tell time to the hour and half-hour (**1.MD.3**). Students construct simple clocks and begin to understand the hour hand, then the minute hand, and then both together. Throughout each lesson, students read both digital and analog clocks to tell time.

Throughout Module 5, students continue daily fluency with addition and subtraction, preparing for Module 6, where they will add within 100 and ensure their mastery of the grade-level fluency goal of sums and differences within 10.

Module Overview 1•5

Notes on Pacing for Differentiation

The work of this module is foundational to the Geometry domain of the Grade 1 standards. Therefore, it is not recommended to omit any lessons from Module 5.

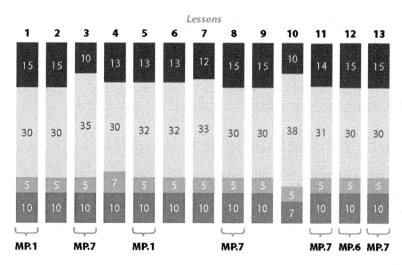

Focus Grade Level Standards

Tell and write time and money.[1]

1.MD.3   Tell and write time in hours and half-hours using analog and digital clocks. Recognize and identify coins, their names, and their values.

Reason with shapes and their attributes.

1.G.1   Distinguish between defining attributes (e.g., triangles are closed and three-sided) versus non-defining attributes (e.g., color, orientation, overall size); build and draw shapes to possess defining attributes.

---

[1] Time alone is addressed in this module. The second portion, regarding money, was added by New York State and is addressed in Module 6.

Module 5:   Identifying, Composing, and Partitioning Shapes

1.G.2   Compose two-dimensional shapes (rectangles, squares, trapezoids, triangles, half-circles, and quarter-circles) or three-dimensional shapes (cubes, right rectangular prisms, right circular cones, and right circular cylinders) to create a composite shape, and compose new shapes from the composite shape. (Students do not need to learn formal names such as "right rectangular prism.")

1.G.3   Partition circles and rectangles into two and four equal shares, describe the shares using the words *halves, fourths*, and *quarters*, and use the phrases *half of, fourth of*, and *quarter of*. Describe the whole as two of, or four of the shares. Understand for these examples that decomposing into more equal shares creates smaller shares.

Foundational Standards

K.G.2   Correctly name shapes regardless of their orientations or overall size.

K.G.3   Identify shapes as two-dimensional (lying in a plane, "flat") or three-dimensional ("solid").

K.G.4   Analyze and compare two- and three-dimensional shapes, in different sizes and orientations, using informal language to describe their similarities, differences, parts (e.g., number of sides and vertices/"corners") and other attributes (e.g., having sides of equal length).

K.G.6   Compose simple shapes to form larger shapes. For example, *"Can you join these two triangles with full sides touching to make a rectangle?"*

Focus Standards for Mathematical Practice

MP.1   **Make sense of problems and persevere in solving them.** Although some students thrive on the visual–spatial perspective of geometric concepts, it can be quite challenging for others. Throughout the module, students are encouraged to continue working toward success when trying to arrange shapes to create specific composite shapes and when recomposing the pieces into different shapes. For some students, sorting shapes into groups without using the common shape names can also create challenges through which they must persevere. This takes place as students distinguish shapes from among variants, palpable distractors, and difficult distractors in Topic A. See examples to the right.[2]

MP.6   **Attend to precision.** Students use clear definitions with peers as they define attributes. For example, while working with a partner, students describe a composite figure by explaining surfaces, sides, and corners so that their partners can create the same composite shape without seeing a visual representation. Students appropriately name parts of a whole using terms such as *halves, fourths,* and *quarters*.

MP.7   **Look for and make use of structure.** Students identify attributes in order to classify shapes such as triangles and cylinders. Students recognize that attributes such as the number of sides, surfaces, etc., are defining attributes, whereas color, size, and orientation are not. Students use their understanding of the partitioning of a circle to tell time.

---

[2]This image, plus further clarification, is found in the Geometry Progressions document, p. 6.

Overview of Module Topics and Lesson Objectives

Standards	Topics and Objectives		Days
1.G.1	A	**Attributes of Shapes** Lesson 1: Classify shapes based on defining attributes using examples, variants, and non-examples. Lesson 2: Find and name two-dimensional shapes including trapezoid, rhombus, and a square as a special rectangle, based on defining attributes of sides and corners. Lesson 3: Find and name three-dimensional shapes including cone and rectangular prism, based on defining attributes of faces and points.	3
1.G.2	B	**Part–Whole Relationships Within Composite Shapes** Lesson 4: Create composite shapes from two-dimensional shapes. Lesson 5: Compose a new shape from composite shapes. Lesson 6: Create a composite shape from three-dimensional shapes and describe the composite shape using shape names and positions.	3
1.G.3	C	**Halves and Quarters of Rectangles and Circles** Lesson 7: Name and count shapes as parts of a whole, recognizing relative sizes of the parts. Lessons 8–9: Partition shapes and identify halves and quarters of circles and rectangles.	3
1.MD.3 1.G.3	D	**Application of Halves to Tell Time** Lesson 10: Construct a paper clock by partitioning a circle and tell time to the hour. Lessons 11–13: Recognize halves within a circular clock face and tell time to the half-hour.	4
		End-of-Module Assessment: Topics A–D (assessment ½ day, return ½ day, remediation or further applications 1 day)	2
Total Number of Instructional Days			**15**

Module 5: Identifying, Composing, and Partitioning Shapes

Terminology

New or Recently Introduced Terms

- Attributes (characteristics of an object such as color or number of sides)
- Composite shapes (shapes composed of two or more shapes)
- Digital clock
- Face (two-dimensional surface of a three-dimensional solid)
- Fourth of (shapes), fourths (1 out of 4 equal parts)
- Half-hour (interval of time lasting 30 minutes)
- Half of, halves (1 out of 2 equal parts)
- Half past (expression for 30 minutes past a given hour)
- Hour (unit for measuring time, equivalent to 60 minutes or 1/24 of a day)
- Hour hand (component on clock tracking hours)
- Minute (unit for measuring time, equivalent to 60 seconds or 1/60 of an hour)
- Minute hand (component on clock tracking minutes)
- O'clock (used to indicate time to a precise hour, with no additional minutes)
- Quarter of (shapes) (1 out of 4 equal parts)
- Three-dimensional shapes:
    - Cone
    - Rectangular prism
- Two-dimensional shapes:
    - Half-circle
    - Quarter-circle
    - Rhombus (flat figure enclosed by four straight sides of the same length wherein two pairs of opposite sides are parallel)
    - Trapezoid (a quadrilateral in which at least one pair of opposite sides is parallel[3])

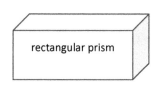

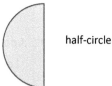
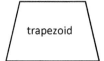

---

[3]This is the formal definition that students learn in Grade 4. It is placed here to signify to teachers the precise definition used in later grades and is not required to be shared with students now. Descriptive explanations such as, "This is a trapezoid. What are its interesting features?" are the general expectation for Grades 1 and 2.

A STORY OF UNITS — Module Overview 1•5

Familiar Terms and Symbols[4]

- Clock
- Shape names (two-dimensional and three-dimensional) from Kindergarten:

  - Circle
  - Cube
  - Cylinder
  - Hexagon (flat figure enclosed by six straight sides)
  - Rectangle (flat figure enclosed by four straight sides and four right angles)
  - Sphere
  - Square (rectangle with four sides of the same length)
  - Triangle (flat figure enclosed by three straight sides)

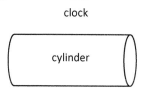
clock

cylinder

sphere

Suggested Tools and Representations

- Pattern blocks
- Square tiles
- Straws
- Student clocks, preferably with gears that can provide the appropriate hour-hand alignment
- Three-dimensional shape models (commercially produced or commonly found examples) including cube, cone, cylinder, rectangular prism, and sphere

Homework

Homework at the K–1 level is not a convention in all schools. In this curriculum, homework is an opportunity for additional practice of the content from the day's lesson. The teacher is encouraged, with the support of parents, administrators, and colleagues, to discern the appropriate use of homework for his or her students. Fluency exercises can also be considered as an alternative homework assignment.

---

[4]These are terms and symbols students have seen previously.

Module 5: Identifying, Composing, and Partitioning Shapes

Scaffolds[5]

The scaffolds integrated into *A Story of Units* give alternatives for how students access information as well as express and demonstrate their learning. Strategically placed margin notes are provided within each lesson elaborating on the use of specific scaffolds at applicable times. They address many needs presented by English language learners, students with disabilities, students performing above grade level, and students performing below grade level. Many of the suggestions are organized by Universal Design for Learning (UDL) principles and are applicable to more than one population. To read more about the approach to differentiated instruction in *A Story of Units,* please refer to "How to Implement *A Story of Units.*"

Assessment Summary

Type	Administered	Format	Standards Addressed
End-of-Module Assessment Task	After Topic D	Constructed response with rubric	1.MD.3 1.G.1 1.G.2 1.G.3

---

[5]Students with disabilities may require Braille, large print, audio, or special digital files. Please visit the website www.p12.nysed.gov/specialed/aim for specific information on how to obtain student materials that satisfy the National Instructional Materials Accessibility Standard (NIMAS) format.

A STORY OF UNITS

GRADE 1

Mathematics Curriculum

GRADE 1 • MODULE 5

Topic A
Attributes of Shapes

1.G.1

Focus Standard:	1.G.1	Distinguish between defining attributes (e.g., triangles are closed and three-sided) versus non-defining attributes (e.g., color, orientation, overall size); build and draw shapes to possess defining attributes.
Instructional Days:	3	
Coherence -Links from:	GK–M2	Two-Dimensional and Three-Dimensional Shapes
-Links to:	G2–M8	Time, Shapes, and Fractions as Equal Parts of Shapes

In Module 5, students build on their exploration and knowledge of shapes from Kindergarten. In Topic A, students identify the defining attributes of individual shapes.

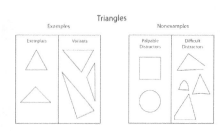

In Lesson 1, students use straws cut at various lengths to create and then classify shapes. A list of the attributes that are common to a set of shapes is created. As students create a new shape with their straws, they decide if it has all the listed attributes. The names of these shapes are intentionally omitted during this lesson to encourage students to use precise language as they describe each shape. In this way, students attend to, and clarify, a shape's defining attributes (1.G.1). For instance, rather than describing a shape as a triangle, students describe it as having three sides and three corners. As students sort the shapes as examples and non-examples, they do the thoughtful work that is depicted in the image to the right at a first-grade level.[1] Students are introduced to the term attributes during this lesson and continue to use the new vocabulary throughout the lessons that follow.

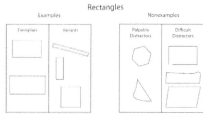

Exemplars are the typical visual prototypes of the shape category.

Variants are other examples of the shape category.

Palpable distractors are nonexamples with little or no overall resemblance to the exemplars.

Difficult distractors are visually similar to examples but lack at least one defining attribute.

In Lesson 2, students connect defining attributes to the classification name. Along with circle, triangle, rectangle, and hexagon, which were introduced in Kindergarten, students learn trapezoid and rhombus. As in Kindergarten, students see squares as special rectangles.

---

[1] This image, plus further clarification, is found in the Geometry Progressions document, p. 6.

Topic A: Attributes of Shapes 9

In Lesson 3, defining attributes of three-dimensional shapes are explored. Along with the three-dimensional shape names learned in Kindergarten (*sphere, cube,* and *cylinder*), students expand their vocabulary to include *cone* and *rectangular prism*. Students are presented with models of three-dimensional shapes as well as real life examples to sort and classify based on defining attributes. Students complete sentence frames that help to distinguish defining attributes from non-defining attributes. For example, "A __[can]__ is in the shape of the __[cylinder]__. It has circles at the ends just like all cylinders. This cylinder is made of metal, but some cylinders are not."

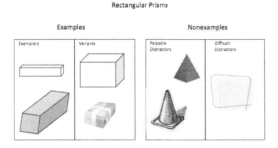

A Teaching Sequence Toward Mastery of Attributes of Shapes	
Objective 1:	Classify shapes based on defining attributes using examples, variants, and non-examples. (Lesson 1)
Objective 2:	Find and name two-dimensional shapes including trapezoid, rhombus, and a square as a special rectangle, based on defining attributes of sides and corners. (Lesson 2)
Objective 3:	Find and name three-dimensional shapes including cone and rectangular prism, based on defining attributes of faces and points. (Lesson 3)

Lesson 1

Objective: Classify shapes based on defining attributes using examples, variants, and non-examples.

Suggested Lesson Structure

■ Fluency Practice   (15 minutes)
■ Application Problem   (5 minutes)
■ Concept Development   (30 minutes)
■ Student Debrief   (10 minutes)
   Total Time   **(60 minutes)**

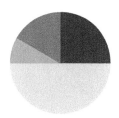

Fluency Practice (15 minutes)

- Grade 1 Core Fluency Sprint   **1.OA.6**   (10 minutes)
- Make it Equal: Addition Expressions   **1.OA.7**   (5 minutes)

Grade 1 Core Fluency Sprint (10 minutes)

Materials: (S) Core Fluency Sprint

Note: For the remainder of the year, a portion of each lesson is devoted to either Core Fluency Sprints or Core Fluency Practice Sets. When Sprints are suggested, choose a Core Fluency Sprint that meets students' needs. All five Core Fluency Sprints are provided at the end of this lesson and are described below for easy reference. Prepare class sets or save the masters for later use because they are not included in future lessons. With each Sprint, notice how many problems most of the class is able to complete; discuss and celebrate improvement as students progress toward Grade 1's required fluency. Quadrants 1, 2, and 3 of each Sprint target Grade 1's core fluency, while Quadrant 4 of the Sprint sometimes extends beyond the grade-level required fluency.

Core Fluency Sprint List:

- Core Addition Sprint 1 (Targets core addition and missing addends.)
- Core Addition Sprint 2 (Targets the most challenging addition within 10.)
- Core Subtraction Sprint (Targets core subtraction.)
- Core Fluency Sprint: Totals of 5, 6, and 7 (Develops understanding of the relationship between addition and subtraction.)
- Core Fluency Sprint: Totals of 8, 9, and 10 (Develops understanding of the relationship between addition and subtraction.)

Make it Equal: Addition Expressions (5 minutes)

Materials: (S) Numeral cards including one "=" card and two "+" cards (Fluency Template)

Note: This activity builds fluency with Grade 1's core addition facts and promotes an understanding of equality.

Assign students partners of equal ability. Students arrange numeral cards from 0 to 10, including the extra 5. Place the "=" card between the partners. Write four numbers on the board (e.g., 9, 5, 5, 1). Partners take the numeral cards that match the numbers written to make two equivalent expressions (e.g., 5 + 5 = 9 + 1).

Suggested sequences: 5, 5, 9, 1; 0, 1, 9, 10; 10, 8, 2, 0; 8, 7, 3, 2; 5, 3, 5, 7; 3, 6, 7, 4; and 2, 4, 6, 8.

Application Problem (5 minutes)

Today, everyone will get 7 straw pieces to use in our lesson. Later, you will use your pieces and your partner's pieces together. How many straw pieces will you have to use when you and your partner put them together?

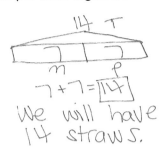

NOTES ON MULTIPLE MEANS OF REPRESENTATION:

Today, students are introduced to the straw kit that is used throughout Module 5. Since students have not worked with the straw pieces, show the class what the straw pieces look like before they begin the Application Problem.

Note: Today's Application Problem is a *put together with total unknown* problem type. Some students may have difficulty determining the second addend since it is not directly stated in the problem. When working with students who are having difficulty, ask these prompting questions: *Can you draw something? What can you draw? What does your drawing show you?* During the Debrief, invite students to explain how they solved the problem.

Concept Development (30 minutes)

Materials: (T) Chart paper, document camera, open- and closed-shape images (Template 1), square corner tester (Template 2) (S) Blank paper, straw kit (see note), ruler

Note: Prepare the square corner tester by cutting out the L shape from the template. Prepare a straw kit for each student. Coffee straws are recommended because they do not roll as easily and fit more neatly on student desks; however, any available straws can be used. Each student kit contains three sizes of straw pieces, created using four straws: 2 full-length straws, 3 half-length straws, and 2 quarter-length straws. The ruler is used for drawing straight lines. Check shape posters and any shape resources that are used to ensure that they show the shapes and names accurately.

A STORY OF UNITS   Lesson 1 1•5

Have students sit at their desks or tables with their materials.

- T: Today, we will be making all kinds of shapes with these straws. Take two minutes to explore the pieces and see what you can make. Keep the straws flat on your desk.
- T: (While students explore and create shapes, circulate and notice how they engage with the materials. Do not discuss shape names with students at this time but rather focus on the number of straight sides, the number of corners, and the length of the sides. During Lesson 2, names are added to the sets of attributes.)
- T: (Project open- and closed-shape images.) Some of you created designs that are open, like this (point to the design labeled *Open Shapes*), and some of you created designs that are closed, like this (point to the design labeled *Closed Shapes*). Think back to what you learned in Kindergarten. Can you remember what the difference is between an open shape (point to the image) and a closed shape (point to the image)?

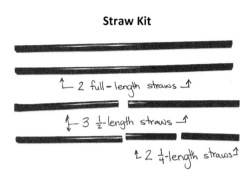

Straw Kit

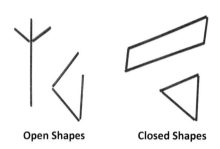

Open Shapes   Closed Shapes

- S: A closed shape is one that has no opening to get out if you were inside the lines. → There's an inside and an outside for a closed shape. → Both ends of every straw touch another straw.
- T: Who has an example of an open shape to show us?
- S: (Share.)
- T: Who has an example of a closed shape to show us?
- S: (Share.)
- T: Today, we'll be making closed shapes, so try to make sure you keep your straws touching at the ends when we make our shapes. If you have an open shape right now, make a new shape so that you have a closed one.
- T: (Look for a student who created a three-sided shape, and place the configuration under the document camera.) Let's look at this shape. How would you describe it?

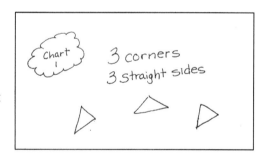

- S: It has three straight sides. → The straws come together at three points. → It has three corners. → The sides are different lengths. (Or, the sides are the same length, depending on the shape displayed.)
- T: (Write *3 Straight Sides* and *3 Corners* at the top of the chart paper.) Use your straws to create this exact same shape on top of your blank paper.
- S: (Create the shape with straws.)

Lesson 1: Classify shapes based on defining attributes using examples, variants, and non-examples.   13

T: (As students do this, ask questions to draw attention to the length of the sides so that students are creating the same exact congruent shape.)

T: Let's record the shape. Draw a dot at the corners where each set of straws meet. Remember a corner is where two sides meet.

T/S: (Draw dots.)

T: (Demonstrate as you describe the process.) Now, move your straws away. Line up your ruler so that two dots are touching the side of the ruler. We can touch one dot with our pencil and draw a very straight line to the next dot. You try it.

S: (Draw a straight line connecting the dots.)

T: Great job! Let's do the same thing to draw all three sides of our shape.

S: (Complete the drawing.)

T: (Put the shape back under the document camera.) Does anyone else have a shape that is made with three straight sides and three corners?

Repeat the process at least four times to create and record various combinations of three straight sides and three corners.

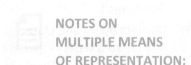

NOTES ON MULTIPLE MEANS OF REPRESENTATION:

Some students find visual discrimination challenging, particularly when the attributes are at more refined levels. Encourage students to persevere. Have students touch the corners or line up the straws as methods to concretely confirm the attributes discussed.

T: (Point to the shapes on the chart.) All of these shapes have two **attributes**, or characteristics, in common. What are they?

S: All of the shapes have three straight sides and three corners.

T: Great! Let's make a new chart with shapes that have a different attribute. Let's make different shapes that all have four straight sides and four corners. Turn over your paper so you can record the shapes on the other side.

Write *4 Corners* and *4 Straight Sides* at the top of a new piece of chart paper. Repeat the process from above at least four times, being sure to include shapes such as two rectangles of varied lengths, a trapezoid, and at least one quadrilateral that is not easily named.

T: Now, combine your straws with your partner. Can you come up with other shapes with four corners and four straight sides that we did not record on our list?

S: (Work with a partner and create shapes such as squares and rhombuses.)

Continue the process of adding these shapes to the chart and having students record the shapes.

T: Let's look at Chart 2. All of these shapes have four straight sides and four corners. Some of the corners are a special kind, called a square corner. They form this shape. (Hold up and trace the edge of the square corner tester.) Let's use the square corner tester to find square corners on these four-sided, four-cornered shapes. (Use the tester, placing a square in the corner of each square corner.)

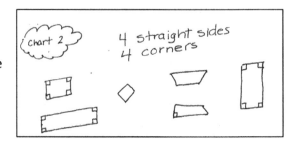

Lesson 1: Classify shapes based on defining attributes using examples, variants, and non-examples.

A STORY OF UNITS   Lesson 1 1•5

T: Think back to the shapes you made earlier. What closed shapes did you make that would *not* fit with one of our charts? We'll make a separate chart for these.

T/S: (Share shapes with five or more straight sides. As students share, create a final chart. Draw each shape, and write its specific attributes next to it.)

T: This paper shows shapes with five straight sides, six straight sides, and even seven straight sides. I want to draw a shape on here that has no straight sides. Who would like to add a shape on here that has no straight sides?

S: (Adds an oval or circle to the chart.)

T: Let's add one open figure, or shape, to the chart as well. (Student adds an open figure.)

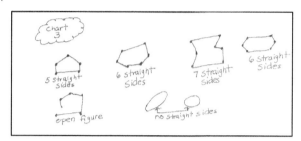

Problem Set (10 minutes)

Students should do their personal best to complete the Problem Set within the allotted 10 minutes.

For some classes, it may be appropriate to modify the assignment by specifying which problems students should work on first. With this option, let the purposeful sequencing of the Problem Set guide the selections so that problems continue to be scaffolded. Balance word problems with other problem types to ensure a range of practice.

Student Debrief (10 minutes)

Lesson Objective: Classify shapes based on defining attributes using examples, variants, and non-examples.

The Student Debrief is intended to invite reflection and active processing of the total lesson experience.

Invite students to review their solutions for the Problem Set. They should check work by comparing answers with a partner before going over answers as a class. Look for misconceptions or misunderstandings that can be addressed in the Debrief. Guide students in a conversation to debrief the Problem Set and process the lesson.

Any combination of the questions below may be used to lead the discussion.

- Look at Problem 1. Which shapes did you choose? Which shapes did not have the **attribute** of having five straight sides?

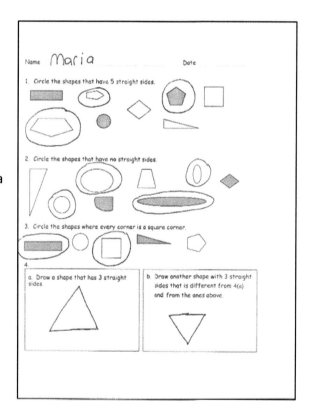

Lesson 1:   Classify shapes based on defining attributes using examples, variants, and non-examples.

15

- Look at Problem 4. Compare your shapes to those on our chart. Which shapes look exactly the same? Did anyone draw a shape that is not already represented on our chart?
- Look at Problem 5. Which attributes, or characteristics, are the same for all of the shapes? Which attributes are different among the shapes in Group A?
- What does it mean to share an attribute of a shape?
- Look at your Application Problem and share your solution with a partner. How did your straws help you create different shapes today? Can you make a shape with four straight sides and only three corners? What would that look like? (Students may put two sides next to each other, essentially making a longer line out of two of the four straws. If this is done, let students know this can be considered one side that uses two straws.)

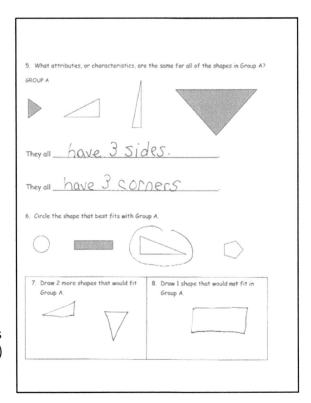

Exit Ticket (3 minutes)

After the Student Debrief, instruct students to complete the Exit Ticket. A review of their work will help with assessing students' understanding of the concepts that were presented in today's lesson and planning more effectively for future lessons. The questions may be read aloud to the students.

Homework

Homework at the K–1 level is not a convention in all schools. In this curriculum, homework is an opportunity for additional practice of the content from the day's lesson. The teacher is encouraged, with the support of parents, administrators, and colleagues, to discern the appropriate use of homework for his or her students. Fluency exercises can also be considered as an alternative homework assignment.

A

Name _____ **Date** _____

Number Correct:

*Write the unknown number. Pay attention to the symbols.

1.	4 + 1 = ____	16.	4 + 3 = ____
2.	4 + 2 = ____	17.	____ + 4 = 7
3.	4 + 3 = ____	18.	7 = ____ + 4
4.	6 + 1 = ____	19.	5 + 4 = ____
5.	6 + 2 = ____	20.	____ + 5 = 9
6.	6 + 3 = ____	21.	9 = ____ + 4
7.	1 + 5 = ____	22.	2 + 7 = ____
8.	2 + 5 = ____	23.	____ + 2 = 9
9.	3 + 5 = ____	24.	9 = ____ + 7
10.	5 + ____ = 8	25.	3 + 6 = ____
11.	8 = 3 + ____	26.	____ + 3 = 9
12.	7 + 2 = ____	27.	9 = ____ + 6
13.	7 + 3 = ____	28.	4 + 4 = ____ + 2
14.	7 + ____ = 10	29.	5 + 4 = ____ + 3
15.	____ + 7 = 10	30.	____ + 7 = 3 + 6

Lesson 1: Classify shapes based on defining attributes using examples, variants, and non-examples.

B Number Correct:

Name _____ Date _____

*Write the unknown number. Pay attention to the symbols.

1.	5 + 1 = ____	16.	2 + 4 = ____
2.	5 + 2 = ____	17.	____ + 4 = 6
3.	5 + 3 = ____	18.	6 = ____ + 4
4.	4 + 1 = ____	19.	3 + 4 = ____
5.	4 + 2 = ____	20.	____ + 3 = 7
6.	4 + 3 = ____	21.	7 = ____ + 4
7.	1 + 3 = ____	22.	4 + 5 = ____
8.	2 + 3 = ____	23.	____ + 4 = 9
9.	3 + 3 = ____	24.	9 = ____ + 5
10.	3 + ____ = 6	25.	2 + 6 = ____
11.	____ + 3 = 6	26.	____ + 6 = 9
12.	5 + 2 = ____	27.	9 = ____ + 2
13.	5 + 3 = ____	28.	3 + 3 = ____ + 4
14.	5 + ____ = 8	29.	3 + 4 = ____ + 5
15.	____ + 3 = 8	30.	____ + 6 = 2 + 7

Lesson 1: Classify shapes based on defining attributes using examples, variants, and non-examples.

A STORY OF UNITS

Lesson 1 Core Addition Sprint 2

A

Name _____

Number Correct: _____

Date _____

*Write the unknown number. Pay attention to the equal sign.

1.	5 + 2 = ____	16.	____ = 5 + 4
2.	6 + 2 = ____	17.	____ = 4 + 5
3.	7 + 2 = ____	18.	6 + 3 = ____
4.	4 + 3 = ____	19.	3 + 6 = ____
5.	5 + 3 = ____	20.	____ = 2 + 6
6.	6 + 3 = ____	21.	2 + 7 = ____
7.	____ = 6 + 2	22.	____ = 3 + 4
8.	____ = 2 + 6	23.	3 + 6 = ____
9.	____ = 7 + 2	24.	____ = 4 + 5
10.	____ = 2 + 7	25.	3 + 4 = ____
11.	____ = 4 + 3	26.	13 + 4 = ____
12.	____ = 3 + 4	27.	3 + 14 = ____
13.	____ = 5 + 3	28.	3 + 6 = ____
14.	____ = 3 + 5	29.	13 + ____ = 19
15.	____ = 3 + 4	30.	19 = ____ + 16

Lesson 1: Classify shapes based on defining attributes using examples, variants, and non-examples.

B

Name _____ Date _____

Number Correct:

*Write the unknown number. Pay attention to the equal sign.

1.	4 + 3 = ____	16.	____ = 6 + 3
2.	5 + 3 = ____	17.	____ = 3 + 6
3.	6 + 3 = ____	18.	5 + 4 = ____
4.	6 + 2 = ____	19.	4 + 5 = ____
5.	7 + 2 = ____	20.	____ = 2 + 7
6.	5 + 4 = ____	21.	2 + 6 = ____
7.	____ = 4 + 3	22.	____ = 3 + 4
8.	____ = 3 + 4	23.	4 + 5 = ____
9.	____ = 5 + 3	24.	____ = 3 + 6
10.	____ = 3 + 5	25.	2 + 7 = ____
11.	____ = 6 + 2	26.	12 + 7 = ____
12.	____ = 2 + 6	27.	2 + 17 = ____
13.	____ = 7 + 2	28.	4 + 5 = ____
14.	____ = 2 + 7	29.	14 + ____ = 19
15.	____ = 7 + 2	30.	19 = ____ + 15

A STORY OF UNITS   Lesson 1 Core Subtraction Sprint   **1•5**

A

Name _____

Number Correct:

Date _____

*Write the unknown number. Pay attention to the symbols.

1.	6 - 1 = ____	16.	8 - 2 = ____
2.	6 - 2 = ____	17.	8 - 6 = ____
3.	6 - 3 = ____	18.	7 - 3 = ____
4.	10 - 1 = ____	19.	7 - 4 = ____
5.	10 - 2 = ____	20.	8 - 4 = ____
6.	10 - 3 = ____	21.	9 - 4 = ____
7.	7 - 2 = ____	22.	9 - 5 = ____
8.	8 - 2 = ____	23.	9 - 6 = ____
9.	9 - 2 = ____	24.	9 - ____ = 6
10.	7 - 3 = ____	25.	9 - ____ = 2
11.	8 - 3 = ____	26.	2 = 8 - ____
12.	10 - 3 = ____	27.	2 = 9 - ____
13.	10 - 4 = ____	28.	10 - 7 = 9 - ____
14.	9 - 4 = ____	29.	9 - 5 = ____ - 3
15.	8 - 4 = ____	30.	____ - 6 = 9 - 7

Lesson 1:   Classify shapes based on defining attributes using examples, variants, and non-examples.

21

©2015 Great Minds. eureka-math.org
G1-M5-TE-BK5-1.3.0-08.2015

A STORY OF UNITS Lesson 1 Core Subtraction Sprint **1•5**

B Number Correct:
Name _____ Date _____

*Write the unknown number. Pay attention to the symbols.

1.	5 - 1 = ____	16.	6 - 2 = ____
2.	5 - 2 = ____	17.	6 - 4 = ____
3.	5 - 3 = ____	18.	8 - 3 = ____
4.	10 - 1 = ____	19.	8 - 5 = ____
5.	10 - 2 = ____	20.	8 - 6 = ____
6.	10 - 3 = ____	21.	9 - 3 = ____
7.	6 - 2 = ____	22.	9 - 6 = ____
8.	7 - 2 = ____	23.	9 - 7 = ____
9.	8 - 2 = ____	24.	9 - ____ = 5
10.	6 - 3 = ____	25.	9 - ____ = 4
11.	7 - 3 = ____	26.	4 = 8 - ____
12.	8 - 3 = ____	27.	4 = 9 - ____
13.	5 - 4 = ____	28.	10 - 8 = 9 - ____
14.	6 - 4 = ____	29.	8 - 6 = ____ - 7
15.	7 - 4 = ____	30.	____ - 4 = 9 - 6

Lesson 1: Classify shapes based on defining attributes using examples, variants, and non-examples.

A STORY OF UNITS — Core Fluency Sprint: Totals of 5, 6, & 7

A

Number Correct:

Name _____   Date _____

*Write the unknown number. Pay attention to the symbols.

1.	2 + 3 =	16.	3 + 3 =
2.	3 + ___ = 5	17.	6 - 3 =
3.	5 - 3 =	18.	6 = ___ + 3
4.	5 - 2 =	19.	2 + 5 =
5.	___ + 2 = 5	20.	5 + ___ = 7
6.	1 + 5 =	21.	7 - 2 =
7.	1 + ___ = 6	22.	7 - 5 =
8.	6 - 1 =	23.	7 = ___ + 5
9.	6 - 5 =	24.	3 + 4 =
10.	___ + 5 = 6	25.	4 + ___ = 7
11.	4 + 2 =	26.	7 - 4 =
12.	2 + ___ = 6	27.	7 = ___ + 3
13.	6 - 2 =	28.	3 = 7 -
14.	6 - 4 =	29.	7 - 5 = ___ - 4
15.	___ + 4 = 6	30.	___ - 3 = 7 - 4

Lesson 1: Classify shapes based on defining attributes using examples, variants, and non-examples.

B

Name _____

Number Correct:

Date _____

*Write the unknown number. Pay attention to the symbols.

1.	1 + 4 =	16.	3 + 3 =
2.	4 + ___ = 5	17.	6 - 3 =
3.	5 - 4 =	18.	6 = ___ + 3
4.	5 - 1 =	19.	2 + 4 =
5.	___ + 1 = 5	20.	4 + ___ = 6
6.	5 + 2 =	21.	6 - 2 =
7.	5 + ___ = 7	22.	6 - 4 =
8.	7 - 2 =	23.	6 = ___ + 4
9.	7 - 5 = ___	24.	3 + 4 =
10.	___ + 2 = 7	25.	4 + ___ = 7
11.	1 + 5 =	26.	7 - 4 =
12.	1 + ___ = 6	27.	7 = ___ + 4
13.	6 - 1 =	28.	4 = 7 - ___
14.	6 - 5 =	29.	6 - 4 = ___ - 5
15.	___ + 5 = 6	30.	___ - 2 = 7 - 3

A Story of Units — Core Fluency Sprint: Totals of 8, 9, & 10

A

Name _____

Number Correct:

Date _____

*Write the unknown number. Pay attention to the symbols.

#	Problem	#	Problem
1.	5 + 5 =	16.	2 + 6 =
2.	5 + ____ = 10	17.	8 = 6 +
3.	10 − 5 =	18.	8 − 2 =
4.	9 + 1 =	19.	2 + 7 =
5.	1 + ____ = 10	20.	9 = 7 +
6.	10 − 1 =	21.	9 − 7 =
7.	10 − 9 =	22.	8 = ____ + 2
8.	___ + 9 = 10	23.	8 − 6 =
9.	1 + 8 =	24.	3 + 6 =
10.	8 + ____ = 9	25.	9 = 6 +
11.	9 − 1 =	26.	9 − 6 =
12.	9 − 8 =	27.	9 = ____ + 3
13.	___ + 1 = 9	28.	3 = 9 −
14.	4 + 4 =	29.	9 − 5 = ____ − 6
15.	8 − 4 =	30.	___ − 7 = 8 − 6

Lesson 1: Classify shapes based on defining attributes using examples, variants, and non-examples.

B

Name _____ Date _____

*Write the unknown number. Pay attention to the symbols.

1.	9 + 1 =	16.	3 + 5 =
2.	1 + ___ = 10	17.	8 = 5 +
3.	10 - 1 =	18.	8 - 3 =
4.	10 - 9 =	19.	2 + 6 =
5.	___ + 9 = 10	20.	8 = 6 +
6.	1 + 7 =	21.	8 - 6 =
7.	7 + ___ = 8	22.	2 + 7 =
8.	8 - 1 =	23.	9 = ___ + 2
9.	8 - 7 =	24.	9 - 7 =
10.	___ + 1 = 8	25.	4 + 5 =
11.	2 + 8 =	26.	9 = 5 +
12.	2 + ___ = 10	27.	9 - 5 =
13.	10 - 2 =	28.	5 = 9 -
14.	10 - 8 =	29.	9 - 6 = ___ - 5
15.	___ + 8 = 10	30.	___ - 6 = 9 - 7

Name _____ Date _____

1. Circle the shapes that have 5 straight sides.

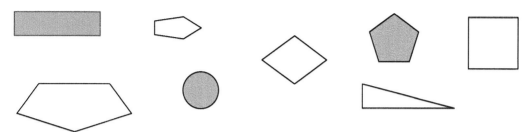

2. Circle the shapes that have no straight sides.

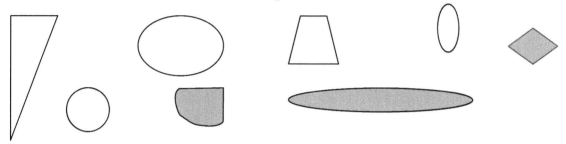

3. Circle the shapes where every corner is a square corner.

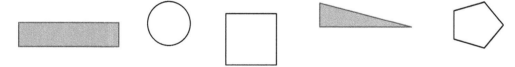

4.
a. Draw a shape that has 3 straight sides.

b. Draw another shape with 3 straight sides that is different from 4(a) and from the ones above.

Lesson 1: Classify shapes based on defining attributes using examples, variants, and non-examples.

5. Which attributes, or characteristics, are the same for all of the shapes in Group A?

GROUP A

They all _____.

They all _____.

6. Circle the shape that best fits with Group A.

  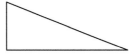

7. Draw 2 more shapes that would fit in Group A.	8. Draw 1 shape that would **not** fit in Group A.

Name _____ Date _____

1. How many corners and straight sides does each of the shapes below have?

a. 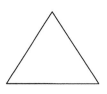 ____ corners ____ straight sides	b. 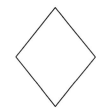 ____ corners ____ straight sides	c. ____ corners ____ straight sides

2. Look at the sides and corners of the shapes in each row.

a. Cross off the shape that does not have the same number of sides and corners.
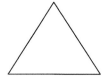 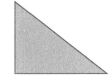   
b. Cross off the shape that does not have the same kind of corners as the other shapes.
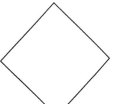

Name _____ Date _____

1. Circle the shapes that have 3 straight sides.

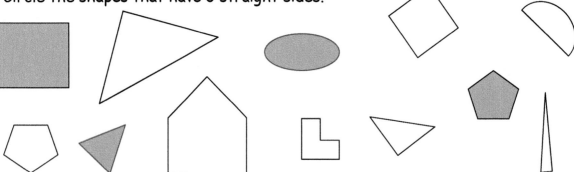

2. Circle the shapes that have no corners.

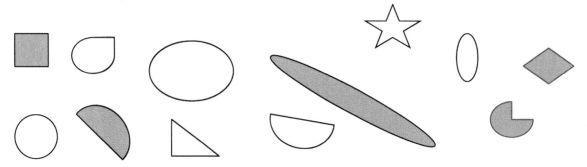

3. Circle the shapes that have only square corners.

4.
a. Draw a shape that has 4 straight sides.

b. Draw another shape with 4 straight sides that is different from 4(a) and from the ones above.

A STORY OF UNITS   Lesson 1 Homework 1•5

5. Which attributes, or characteristics, are the same for all of the shapes in Group A?

GROUP A

    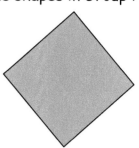

They all _____.

They all _____.

6. Circle the shape that best fits with Group A.

7. Draw 2 more shapes that would fit in Group A.	8. Draw 1 shape that would **not** fit in Group A.

A STORY OF UNITS

Lesson 1 Fluency Template 1•5

0	1	2	3
4	5	<u>6</u>	7
8	<u>9</u>	10	5
=	+	+	−
−			

numeral cards

Lesson 1: Classify shapes based on defining attributes using examples, variants, and non-examples.

open- and closed-shape images

Print on cardstock, and cut out each of the two square corner testers.

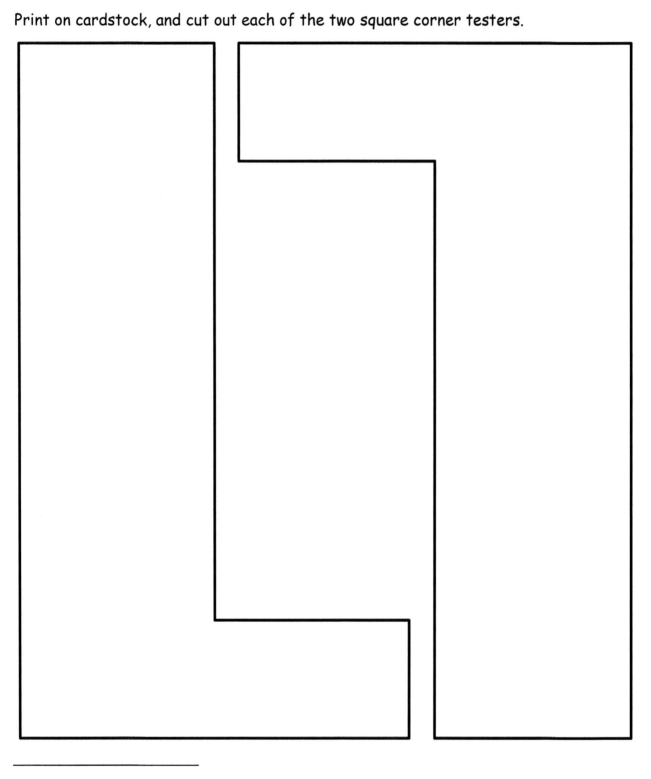

square corner tester

Lesson 2

Objective: Find and name two-dimensional shapes including trapezoid, rhombus, and a square as a special rectangle, based on defining attributes of sides and corners.

Suggested Lesson Structure

- ■ Fluency Practice   (15 minutes)
- ▨ Application Problem   (5 minutes)
- ▨ Concept Development   (30 minutes)
- ■ Student Debrief   (10 minutes)

   Total Time   **(60 minutes)**

Fluency Practice (15 minutes)

- Grade 1 Core Fluency Sprint **1.OA.6**   (10 minutes)
- Make It Equal: Subtraction Expressions **1.OA.7**   (5 minutes)

Grade 1 Core Fluency Sprint (10 minutes)

Materials: (S) Core Fluency Sprint (Lesson 1 Core Fluency Sprint)

Note: Based on the needs of the class, select a Sprint from Lesson 1. Consider the options below:

1. Re-administer the previous lesson's Sprint.
2. Administer the next Sprint in the sequence.
3. Differentiate. Administer two different Sprints. Simply have one group do a counting activity on the back of the Sprint while the other group corrects the second Sprint.

Make It Equal: Subtraction Expressions (5 minutes)

Materials: (S) Numeral cards (Lesson 1 Fluency Template), one "=" card, two "–" cards

Note: This activity builds fluency with subtraction within 10 and promotes an understanding of equality.

Assign students partners of similar skill or ability level. Students arrange numeral cards from 0 to 10, including the extra 5. Place the "=" card between the partners. Write four numbers on the board (e.g., 9, 10, 2, 1). Partners take the numeral cards that match the numbers written to make two equivalent subtraction expressions (e.g., 10 – 9 = 2 – 1). Students can be encouraged to make another sentence of equivalent expressions for the same set of cards as well (e.g., 10 – 2 = 9 – 1). Encourage students to find examples that result in an answer other than 1 = 1, as in the previous example.

Suggested sequence: 10, 9, 2, 1; 2, 10, 3, 9; 4, 5, 9, 10; 10, 8, 7, 9; 7, 10, 9, 6; and 2, 4, 10, 8.

A STORY OF UNITS

Lesson 2 1•5

Application Problem (5 minutes)

Lee has 9 straws. He uses 4 straws to make a shape. How many straws does he have left to make other shapes?

Extension: What possible shapes could Lee have created? Draw the different shapes Lee might have made using 4 straws. Label any shapes whose name you know.

Note: Today's Application Problem uses a familiar context that was established during Lesson 1 of the module. Through the extension, students have the opportunity to apply the previous lesson and generate prior knowledge that is useful for today's objective.

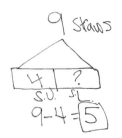

Lee has 5 straws left.

Concept Development (30 minutes)

Materials: (T) Charts from Lesson 1, shape description cards (Template), tape (S) Straw kit, 10 additional straws per person, square corner tester (Lesson 1 Template 2), shape description cards (Template)

Note: The description of each shape is consistent with mathematical descriptions used throughout the K–12 continuum of this curriculum. Below are some clarifying comments about each shape mentioned in this lesson.

Triangle: A triangle can be described based on its three sides or its three corners or angles.

Rectangle: A rectangle is a quadrilateral with four right angles. The length of each side is not a defining attribute. For this reason, a square is a type of rectangle. While some rectangles have two short sides and two longer sides, that is not a requirement or defining attribute of a rectangle.

Rhombus: A rhombus is a quadrilateral with four sides of the same length. The definition does not depend on the measure of its angles. For this reason, a square is also a special type of rhombus that has right angles.

Square: A square is a special shape that is both a rectangle and a rhombus since it is a quadrilateral with four right angles and four sides of the same length.

NOTES ON MULTIPLE MEANS OF REPRESENTATION:

Highlight the critical vocabulary for English language learners throughout the lesson. Key vocabulary words—*characteristic* and *attribute*—were introduced in Lesson 1. Without understanding these words, English language learners will struggle with the first few lessons of this module. Spend some extra time relating the words while describing the classroom or students so that students see the relationship between describing shapes and other things in their environment.

---

36   Lesson 2:   Find and name two-dimensional shapes including trapezoid, rhombus, and a square as a special rectangle, based on defining attributes of sides and corners.

T: Yesterday, you made all of these shapes with your straws. (Show the charts from Lesson 1.) Today, we're going to name them based on their attributes, or characteristics. (Hold up the triangle card.) The word *triangle* actually describes something about the shape! Listen carefully—*tri* means three, and *angle* is what gives us corners. So, when we say *triangle,* we're saying it has three angles, or three corners. Which can we label as triangles?

S: The ones on the first chart. (Students point to the triangles.)

T: Are they all triangles? Tell me about each one.

Students explain or touch each of the three corners of each shape to confirm that they are all triangles. Ensure that students point out that all the triangles also have three straight sides. Tape the triangle description card under the triangles.

T: Let's try another card. (Hold up the hexagon card.) A **hexagon** is a shape with six straight sides. Do we have any hexagons on our chart?

S: (Point to the two hexagons on Chart 3.) Yes, these shapes have six straight sides!

T: (Tape the card on the chart near the hexagons.) Do we have any other hexagons on these charts?

S: No!

Move to the rectangle and square description cards.

T: A rectangle is a shape with four square corners, or right angles. Do we have any rectangles on our chart? Use your square corner tester to check.

S: (Point to any rectangles on the charts, and explain why they fit the description.)

T: (Ensure that students include the squares as shapes that fit the description. Add the rectangle cards under the shapes.) Do any of these rectangles have another name you know?

S: Yes! The square.

T: Yes, a square is a type of special rectangle with four straight sides of equal length. (Tape a square card under the rectangle card.)

T: A **rhombus** is a shape with four straight sides of equal length. Do we have any rhombuses?

S: (Point to shapes with four straight sides of equal length, including the shape that is already labeled with *square* and *rectangle*.)

T: (As students explain how each shape fits the description, tape the description card below the drawing.) Yes, a square is a special kind of rectangle, and it is also a special kind of rhombus. Squares are pretty special!

T: (Point to the example of a trapezoid on the chart.) Does anyone know what this shape is called?

S: A **trapezoid**. (If no one knows the name, tell students it is a trapezoid.)

T: How is this shape the same as the other shapes we have defined?

S: It has four straight sides and four corners.

T: How is this trapezoid different from the other shapes?

S: The sides are not all the same length, like the square. → This trapezoid doesn't have four square corners.

Lesson 2: Find and name two-dimensional shapes including trapezoid, rhombus, and a square as a special rectangle, based on defining attributes of sides and corners.

A STORY OF UNITS

Lesson 2 1•5

T: Now, you're ready to play Make the Shape with your partner. Here's how to play:
- Each pair gets a stack of shape description cards and places 10 additional straws in their straw kit.
- Turn over a card. Use your straws to make that shape, and put the card below your shape.
- Take turns until one player has used all of his straws.
- If you have more time, shuffle up the cards, and take turns trying to pick the cards that match the shapes you've made.

Problem Set (10 minutes)

Students should do their personal best to complete the Problem Set within the allotted 10 minutes. For some classes, it may be appropriate to modify the assignment by specifying which problems they work on first.

Student Debrief (10 minutes)

Lesson Objective: Find and name two-dimensional shapes including trapezoid, rhombus, and a square as a special rectangle, based on defining attributes of sides and corners.

The Student Debrief is intended to invite reflection and active processing of the total lesson experience.

Invite students to review their solutions for the Problem Set. They should check work by comparing answers with a partner before going over answers as a class. Look for misconceptions or misunderstandings that can be addressed in the Debrief. Guide students in a conversation to debrief the Problem Set and process the lesson.

Any combination of the questions below may be used to lead the discussion.
- Look at Problem 1. Which shapes were the most challenging to count or find? Which shapes were the easiest? Explain your thinking.

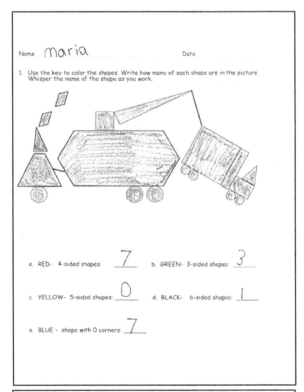

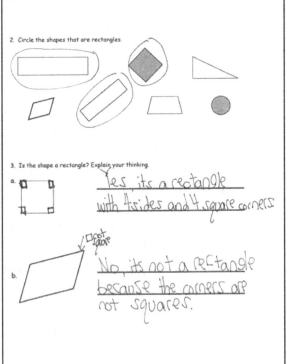

Lesson 2: Find and name two-dimensional shapes including trapezoid, rhombus, and a square as a special rectangle, based on defining attributes of sides and corners.

- Which four-sided shapes are squares? Which are **rhombuses**? Which are rectangles? Which are **trapezoids**? (Note that a square is a type of rectangle *and* a type of rhombus.) How many sides do **hexagons** have?
- What name can we use for the three-sided shapes? What name can we use for the six-sided shapes? What name can we use for all of the shapes with no corners in this picture?
- In Problem 1, what do the shapes look like when they are put together in this way?
- Look at Problem 2. Explain why you chose each shape that is a rectangle. Explain why the other shapes are *not* rectangles.
- Look at Problem 3(b). How is the shape in Problem 3(b) like a rectangle? How is it different from a rectangle? What other shapes have similar attributes to the shape in Problem 3(b)? How are they similar, and how are they different? Explain your thinking.
- What shape names did we use today? Name the attributes or characteristics that are important to each shape.
- Look at the Application Problem. What shape or shapes might Lee have created?
- How did your fluency work go today? How do you practice?

Exit Ticket (3 minutes)

After the Student Debrief, instruct students to complete the Exit Ticket. A review of their work will help with assessing students' understanding of the concepts that were presented in today's lesson and planning more effectively for future lessons. The questions may be read aloud to the students.

A STORY OF UNITS — Lesson 2 Problem Set 1•5

Name _____ Date _____

1. Use the key to color the shapes. Write how many of each shape are in the picture. Whisper the name of the shape as you work.

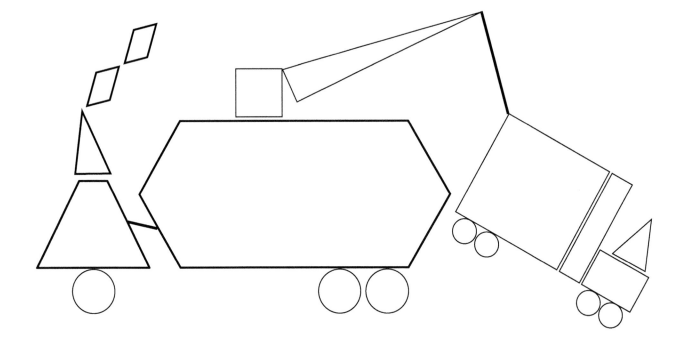

a. RED—4-sided shapes: _____

b. GREEN—3-sided shapes: _____

c. YELLOW—5-sided shapes: _____

d. BLACK—6-sided shapes: _____

e. BLUE—shapes with no corners: _____

2. Circle the shapes that are rectangles.

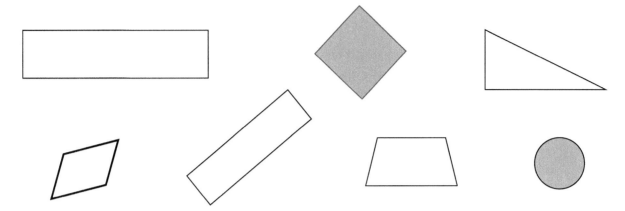

3. Is the shape a rectangle? Explain your thinking.

a.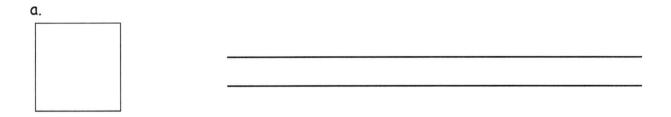

b.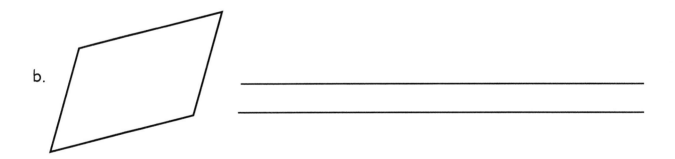

A STORY OF UNITS   Lesson 2 Exit Ticket 1•5

Name _____   Date _____

Write the number of corners and sides that each shape has. Then, match the shape to its name. Remember that some special shapes may have more than one name.

1. 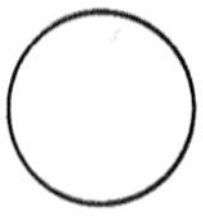 ○
   ____ corners
   ____ straight sides

   triangle

2.  ▽
   ____ corners
   ____ straight sides

   circle

3.  ⬡
   ____ corners
   ____ straight sides

   rectangle

   hexagon

4. ☐
   ____ corners
   ____ straight sides

   square

   rhombus

42   Lesson 2:   Find and name two-dimensional shapes including trapezoid, rhombus, and a square as a special rectangle, based on defining attributes of sides and corners.

Name _____ Date _____

1. Color the shapes using the key. Write the number of shapes you colored on each line.

Key

RED 3 straight sides: _____

BLUE 4 straight sides: _____

GREEN 6 straight sides: _____

YELLOW 0 straight sides: _____

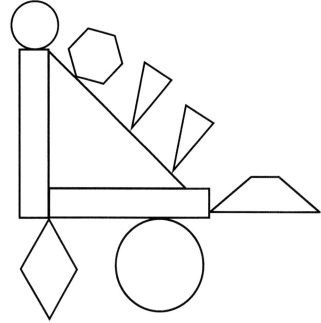

2.
   a. A **triangle** has ____ straight sides and ____ corners.
   b. I colored ____ triangles.

3.
   a. A **hexagon** has ____ straight sides and ____ corners.
   b. I colored ____ hexagon.

4.
   a. A **circle** has ____ straight sides and ____ corners.
   b. I colored ____ circles.

Lesson 2: Find and name two-dimensional shapes including trapezoid, rhombus, and a square as a special rectangle, based on defining attributes of sides and corners.

5.

   a. A **rhombus** has ____ straight sides that are equal in length and ____ corners.

   b. I colored ____ rhombus.

6. A **rectangle** is a closed shape with 4 straight sides and 4 square corners.

   a. Cross off the shape that is NOT a rectangle.

   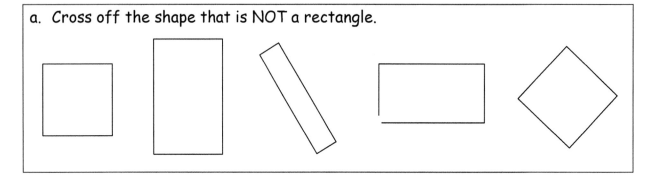

   b. Explain your thinking: _____
   _____

7. A **rhombus** is a closed shape with 4 straight sides of the same length.

   a. Cross off the shape that is NOT a rhombus.

   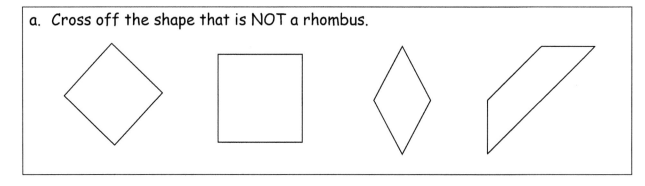

   b. Explain your thinking: _____
   _____

hexagon
closed shape with 6 straight sides

rectangle
closed shape with 4 straight sides and 4 square corners

square
closed shape with 4 straight sides of the same length and 4 square corners

triangle
closed shape with 3 straight sides

rhombus
closed shape with 4 straight sides of the same length

shape description cards

Lesson 2: Find and name two-dimensional shapes including trapezoid, rhombus, and a square as a special rectangle, based on defining attributes of sides and corners.

Lesson 3

Objective: Find and name three-dimensional shapes including cone and rectangular prism, based on defining attributes of faces and points.

Suggested Lesson Structure

■ Fluency Practice   (10 minutes)
■ Application Problem   (5 minutes)
■ Concept Development   (35 minutes)
■ Student Debrief   (10 minutes)
Total Time   **(60 minutes)**

Fluency Practice (10 minutes)

- Grade 1 Core Fluency Differentiated Practice Sets **1.OA.6**   (5 minutes)
- Count by 10 or 1 with Dimes and Pennies **1.NBT.5, 1.MD.3**   (5 minutes)

Grade 1 Core Fluency Differentiated Practice Sets (5 minutes)

Materials: (S) Core Fluency Practice Sets

Note: This activity assesses students' progress toward mastery of the required addition fluency for Grade 1 students. Give the appropriate Practice Set to each student. Students who completed all of the questions correctly on their most recent Practice Set should be given the next level of difficulty. All other students should try to improve their scores on their current level.

Students complete as many problems as they can in 90 seconds. Assign a counting pattern and start number for early finishers, or tell them to practice make ten addition and subtraction on the back of their papers. When time runs out, collect and correct any Practice Sets that are completed.

Count by 10 or 1 with Dimes and Pennies (5 minutes)

Materials: (T) 10 dimes and 10 pennies

Note: This fluency activity uses dimes and pennies as abstract representations of tens and ones to help students become familiar with coins while simultaneously providing practice with counting forward and backward by 10 or 1.

- First minute: Place and take away dimes in a 5-group formation as students count along by 10.
- Second minute: Begin with 2 pennies. Ask how many ones there are. Instruct students to start at 2 and add or subtract 10 while placing and taking away dimes.

- Third minute: Begin with 2 dimes. Ask how many tens there are. Instruct students to begin at 20 and add or subtract 1 while placing and taking away pennies.

Application Problem (5 minutes)

Rose draws 6 triangles. Maria draws 7 triangles. How many more triangles does Maria have than Rose?

Note: Let students know that today's problem is a little different from past problems because today they are comparing Rose's triangles with Maria's. Suggest that they draw two different tapes with the same endpoint on the left, so that they can more easily compare the two numbers. While circulating, support students in aligning their shapes and bars to assist in solving this *compare with difference unknown* problem type.

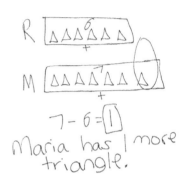

Concept Development (35 minutes)

Materials: (T) Set of three-dimensional shapes (sphere, cone, cube, rectangular prism, and cylinder), three-dimensional shapes found around home or school, three-dimensional shape description cards (Template), tape

Note: If a kit of three-dimensional shapes is not readily available, then a kit can be borrowed from other grade levels, such as Kindergarten (used in Kindergarten Module 2 and Kindergarten Module 6). Another option is to collect three-dimensional shapes from students' homes as suggested below.

- Spheres: balls (e.g., tennis balls) and marbles
- Cylinders: paper towel and oatmeal containers
- Cubes: small tissue boxes, gift boxes, and large dice
- Rectangular prisms: large tissue boxes, crayon boxes, marker boxes, and pencil holders
- Cones: ice cream cones and party hats

NOTES ON MULTIPLE MEANS OF REPRESENTATION:

Be sure to have a pictorial word wall in the classroom that is easily accessible for students. The wall should include the following words at this point in the module: *circle, hexagon, rectangle, rhombus, square, trapezoid, triangle, cone, cube, cylinder, rectangular prism,* and *sphere*. Spending some time learning these words would be helpful to all students, especially the word *cylinder*, whose spelling can be confusing. Also include the describing attributes for three-dimensional solids including *face, edge,* and *vertex*.

Before the lesson, place examples of three-dimensional figures around the room. Gather students in the meeting area in a semicircle.

T: (Place one example of each three-dimensional shape on the floor.) Today, we are going to talk about three-dimensional shapes, like these. What do you know about three-dimensional shapes?

S: They are not flat. → They have different faces or surfaces. → They are solid. → That one is called a cube. (Points to the cube.) → You can touch them on different sides.

T: Great! Yes, three-dimensional shapes have **faces** (touch each face on a cube), and they have different types of corners or points (touch the vertices). Often they are solid and can be called **three-dimensional solids.** There are lots of three-dimensional shapes around our room. Some look just like the materials we have here, and some look different. Can anyone think of an item in the room that looks like these?

S: Our party hat on the teddy bear looks like that one. (Points to the cone.) → That one looks like our dice. (Points to the cube.) → That one looks like the container for our alphabet game! (Points to the cylinder.)

T: Find one item in the room that is three-dimensional—an object that has faces, not a flat two-dimensional shape. You have 30 seconds. Walk, find your item, and bring it to the carpet.

S: (Search the room, and bring back one item each to the carpet.)

T: Someone told us the name of this shape earlier. Who remembers the name of this shape?

S: A cube! (Place the cube in the middle of the meeting area.)

T: What are the attributes, or characteristics, that make this a cube?

S: It has six faces, and every face is a square. (Ask the student to demonstrate this using the cube, and then tape the appropriate shape description card to the cube.)

T: (Place the cube on the carpet.) Let's count the faces of the cube. Track the number with your fingers. The bottom. How many faces is that?

S: One!

T: The top. How many now?

S: Two!

T: Now, let's go around the cube.

S: The side closest to me. How many is that?

S: Three!

T: The side to its right?

S: Four!

Keep going around systematically. Count again to increase students' proficiency.

T: Look at your items. Who brought a cube to the meeting area?

S: (Students show their items.)

T: Let's check. Count the faces of the cube with your partner. (Pause.) Does your cube have six faces?

S: (Count the faces.) Yes.

T: Are all six faces squares?

S: Yes.

Note: A cube is a special type of rectangular prism. On the Problem Set, some students will not notice that the die could also be considered a rectangular prism. As students are ready for this increased complexity, this can be discussed during the Debrief.

Repeat this process with students who believe they have a cube. Some students will answer no to one or both of the questions. Explain that the item must have both attributes to be a cube. If they answer yes to one of the two questions, discuss how the object is like a cube in one way but unlike a cube in another way.

A STORY OF UNITS   Lesson 3   1•5

MP.7
T: How are all of these cubes alike?
S: They all have six square faces.
T: How are they different from each other?
S: Some of them are made of paper. → One of them is made of plastic. → That one is yellow. → The tissue box is empty on the inside, but the dice are not.
T: (Hold up the rectangular prism.) This is a **rectangular prism**. A rectangular prism also has six faces, but let's check. Does it have six faces? (Count with students.)
S: Yes.
T: What shape are the faces?
S: They are all rectangles. → Some faces are squares, but all squares are also special types of rectangles.
T: The attributes of a rectangular prism are that they have six faces, and all of the faces are rectangles. Remember, squares are a special kind of rectangle, so some of your faces might be squares. Who has a rectangular prism in front of them?

NOTES ON MULTIPLE MEANS OF ACTION AND EXPRESSION:

Students may need some extra practice identifying shapes correctly based on attributes. Listening to others talk about shapes helps these students, especially English language learners, understand and acquire language pertaining to this topic.

Like the process of checking each cube, repeat this process with students who believe they have a rectangular prism. If they answer yes to one of the two questions, discuss how the object is like a rectangular prism in one way but unlike a rectangular prism in another way. Ask students which attributes are common to all of the objects and which attributes are found only on some of the objects.

Repeat the process with a cylinder (one circular or oval face or space on each end and one curved side), a **cone** (one circular or oval face or space and one curved side that comes to a point at the other end), and a sphere (one curved surface with no flat faces).

Problem Set (10 minutes)

Students should do their personal best to complete the Problem Set within the allotted 10 minutes. For some classes, it may be appropriate to modify the assignment by specifying which problems they work on first.

Students may or may not notice that the die is considered a cube and a rectangular prism. Challenge students who are ready to find the shape that could be called by two names.

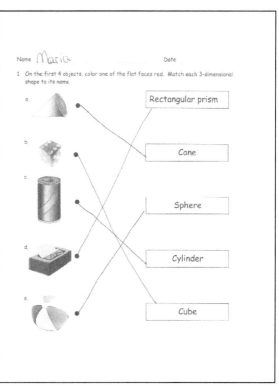

Lesson 3: Find and name three-dimensional shapes including cone and rectangular prism, based on defining attributes of faces and points.

49

Student Debrief (10 minutes)

Lesson Objective: Find and name three-dimensional shapes including cone and rectangular prism, based on defining attributes of faces and points.

The Student Debrief is intended to invite reflection and active processing of the total lesson experience. Invite students to review their solutions for the Problem Set. They should check work by comparing answers with a partner before going over answers as a class. Look for misconceptions or misunderstandings that can be addressed in the Debrief. Guide students in a conversation to debrief the Problem Set and process the lesson.

Any combination of the questions below may be used to lead the discussion.

- Look at Problem 1. Which face did you color on each three-dimensional shape? How did coloring the face help you find the matching shape name?
- Look at Problem 2. Which materials from around the room could you add to each column on the chart? How are the items that are all spheres similar to each other? How are they different? Which attribute is the most important for naming the objects as spheres? (Repeat with each shape.)
- How are the party hat and paper towel roll different from the cylinder and cone in our three-dimensional shapes?
- What are the names of the three-dimensional shapes that we used today? Tell your partner the important attributes of each shape. (Cubes, spheres, **cones**, **rectangular prisms**, and cylinders.)
- Look at your Application Problem. How did you solve this problem? Share drawings and strategies for solving each question.

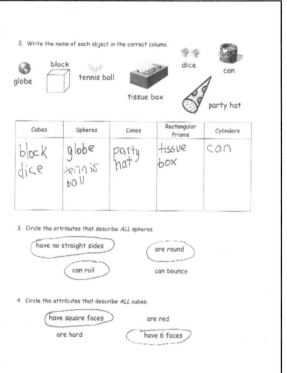

- Think about today's Fluency Practice. What part of today's fluency activities is easier for you now than when we first learned about it? Explain what is easier for you now.

Exit Ticket (3 minutes)

After the Student Debrief, instruct students to complete the Exit Ticket. A review of their work will help with assessing students' understanding of the concepts that were presented in today's lesson and planning more effectively for future lessons. The questions may be read aloud to the students.

Name _____ Date _____

My Addition Practice

1. 6 + 0 = ___
2. 0 + 6 = ___
3. 5 + 1 = ___
4. 1 + 5 = ___
5. 6 + 1 = ___
6. 1 + 6 = ___
7. 6 + 2 = ___
8. 5 + 2 = ___
9. 2 + 5 = ___
10. 2 + 4 = ___
11. 7 + 1 = ___
12. ___ = 1 + 7
13. 3 + 3 = ___
14. 3 + 4 = ___
15. ___ = 3 + 5
16. 6 + 3 = ___
17. 7 + 3 = ___
18. ___ = 7 + 2
19. 2 + 7 = ___
20. 2 + 8 = ___
21. 5 + 3 = ___
22. ___ = 5 + 4
23. 6 + 4 = ___
24. 4 + 6 = ___
25. ___ = 4 + 4
26. 3 + 4 = ___
27. 5 + 5 = ___
28. ___ = 4 + 5
29. 3 + 7 = ___
30. ___ = 3 + 6

Today, I finished _____ problems.

Name _____ Date _____

My Missing Addend Practice

1. 6 + ___ = 6
2. 0 + ___ = 6
3. 5 + ___ = 6
4. 4 + ___ = 6
5. 0 + ___ = 7
6. 6 + ___ = 7
7. 1 + ___ = 7
8. 7 + ___ = 8
9. 1 + ___ = 8
10. 6 + ___ = 8
11. 3 + ___ = 6
12. 4 + ___ = 8
13. 10 = 5 + ___
14. 5 + ___ = 9
15. 5 + ___ = 7
16. 8 = 5 + ___
17. 5 + ___ = 9
18. 8 + ___ = 10
19. 7 + ___ = 10
20. 10 = 6 + ___
21. 4 + ___ = 7
22. 7 = 3 + ___
23. 2 + ___ = 7
24. 2 + ___ = 8
25. 9 = 2 + ___
26. 2 + ___ = 10
27. 10 = 3 + ___
28. 3 + ___ = 9
29. 4 + ___ = 9
30. 10 = 4 + ___

Today, I finished _____ problems.

I solved _____ problems correctly.

Name _____ Date _____

My Related Addition and Subtraction Practice

1. 5 + ___ = 6
2. 1 + ___ = 6
3. 6 - 1 = ___
4. 9 + ___ = 10
5. 1 + ___ = 10
6. 10 - 9 = ___
7. 5 + ___ = 10
8. 10 - 5 = ___
9. 8 + ___ = 10
10. 10 - 8 = ___
11. 7 + ___ = 10
12. 10 - 7 = ___
13. 5 + ___ = 7
14. 7 - 5 = ___
15. 5 + ___ = 8
16. 8 - 5 = ___
17. 4 + ___ = 6
18. 6 - 4 = ___
19. 3 + ___ = 6
20. 6 - 3 = ___
21. 4 + ___ = 8
22. 8 - 4 = ___
23. 4 + ___ = 7
24. 7 - 4 = ___
25. 5 + ___ = 9
26. 9 - 5 = ___
27. 6 + ___ = 9
28. 9 - 6 = ___
29. 4 + ___ = 7
30. 7 - 4 = ___

Today, I finished _____ problems.

I solved _____ problems correctly.

Name _____ Date _____

My Subtraction Practice

1. 6 - 0 = ___
2. 6 - 1 = ___
3. 7 - 1 = ___
4. 8 - 1 = ___
5. 6 - 2 = ___
6. 7 - 2 = ___
7. 9 - 2 = ___
8. 10 - 10 = ___
9. 10 - 9 = ___
10. 10 - 7 = ___
11. 6 - 3 = ___
12. 7 - 3 = ___
13. 9 - 3 = ___
14. 10 - 8 = ___
15. 10 - 6 = ___
16. 10 - 4 = ___
17. 10 - 5 = ___
18. 7 - 6 = ___
19. 7 - 5 = ___
20. 6 - 4 = ___
21. 8 - 4 = ___
22. 8 - 3 = ___
23. 8 - 5 = ___
24. 9 - 5 = ___
25. 9 - 4 = ___
26. 7 - 3 = ___
27. 10 - 7 = ___
28. 9 - 7 = ___
29. 9 - 6 = ___
30. 8 - 6 = ___

Today, I finished _____ problems.

I solved _____ problems correctly.

Name _____ Date _____

My Mixed Practice

1. 4 + 2 = ___
2. 2 + ___ = 6
3. 6 = 3 + ___
4. 2 + 5 = ___
5. 7 = 5 + ___
6. 4 + 3 = ___
7. 7 = ___ + 4
8. 8 = ___ + 4
9. 4 + 5 = ___
10. 9 = ___ + 4

11. 2 + ___ = 6
12. 6 - 2 = ___
13. 6 - 4 = ___
14. 5 + ___ = 7
15. 7 - 5 = ___
16. 7 - 4 = ___
17. 7 - 3 = ___
18. 8 = 6 + ___
19. 8 - 2 = ___
20. 8 - 6 = ___

21. 8 - 5 = ___
22. 3 + ___ = 8
23. 8 = ___ + 5
24. ___ + 2 = 9
25. 9 = ___ + 7
26. 9 - 2 = ___
27. 9 - 7 = ___
28. 9 - 6 = ___
29. 9 = ___ + 4
30. 9 - 6 = ___

Today, I finished _____ problems.

I solved _____ problems correctly.

A STORY OF UNITS

Lesson 3 Problem Set 1•5

Name _____   Date _____

1. On the first 4 objects, color one of the flat faces red. Match each 3-dimensional shape to its name.

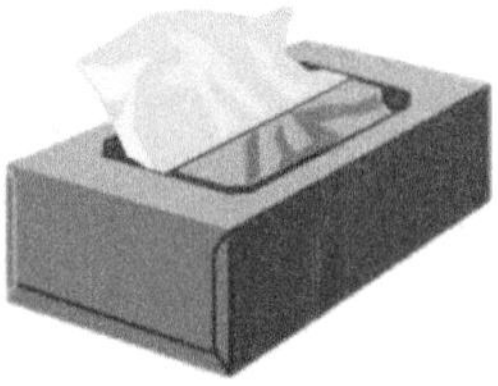

a.

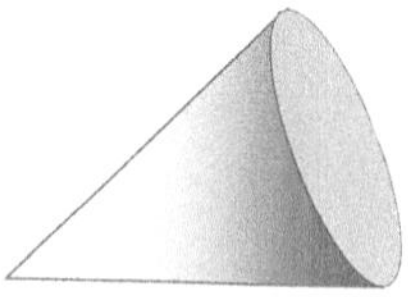

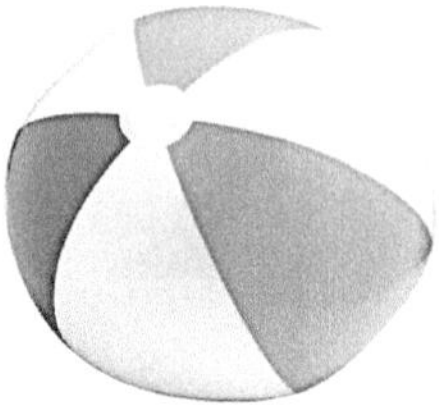

Rectangular prism

b.

Cone

c.

Sphere

d.

Cylinder

e.

Cube

2. Write the name of each object in the correct column.

block

tissue box

dice

can

globe

tennis ball

party hat

Cubes	Spheres	Cones	Rectangular Prisms	Cylinders

3. Circle the attributes that describe ALL spheres.

have no straight sides

are round

can roll

can bounce

4. Circle the attributes that describe ALL cubes.

have square faces

are red

are hard

have 6 faces

Lesson 3: Find and name three-dimensional shapes including cone and rectangular prism, based on defining attributes of faces and points.

Name _____ Date _____

Circle true or false. Write one sentence to explain your answer. Use the word bank if needed.

Word Bank

| faces | circle | square |
| sides | rectangle | point |

1.

This can is a cylinder.   True or False

2.

This juice box is a cube.   True or False

Lesson 3: Find and name three-dimensional shapes including cone and rectangular prism, based on defining attributes of faces and points.

A STORY OF UNITS

Lesson 3 Homework 1•5

Name _____ Date _____

1. Go on a scavenger hunt for 3-dimensional shapes. Look for objects at home that would fit in the chart below. Try to find at least four objects for each shape.

Cube	Rectangular Prism	Cylinder	Sphere	Cone

Lesson 3: Find and name three-dimensional shapes including cone and rectangular prism, based on defining attributes of faces and points.

59

A STORY OF UNITS

Lesson 3 Homework 1•5

2. Choose one object from each column. Explain how you know that object belongs in that column. Use the word bank if needed.

Word Bank

| faces | circle | square | roll | six |
| sides | rectangle | point | flat | |

a. I put the _____ in the cube column because
_____.

b. I put the _____ in the cylinder column because
_____.

c. I put the _____ in the sphere column because
_____.

d. I put the _____ in the cone column because
_____.

e. I put the _____ in the rectangular prism column because _____.

Lesson 3: Find and name three-dimensional shapes including cone and rectangular prism, based on defining attributes of faces and points.

cone
3-dimensional shape with only one circle or oval face and one point

cube
3-dimensional shape with 6 square faces

cylinder
3-dimensional shape with 2 circle or oval faces that are the same size

rectangular prism
3-dimensional shape with 6 rectangle faces

sphere
3-dimensional shape with no flat faces

three-dimensional shape description cards

Mathematics Curriculum

GRADE 1 • MODULE 5

Topic B
Part–Whole Relationships Within Composite Shapes

1.G.2

Focus Standard:	1.G.2	Compose two-dimensional shapes (rectangles, squares, trapezoids, triangles, half-circles, and quarter-circles) or three-dimensional shapes (cubes, right rectangular prisms, right circular cones, and right circular cylinders) to create a composite shape, and compose new shapes from the composite shape. (Students do not need to learn formal names such as "right rectangular prism.")
Instructional Days:	3	
Coherence -Links from:	GK–M2	Two-Dimensional and Three-Dimensional Shapes
-Links to:	G2–M8	Time, Shapes, and Fractions as Equal Parts of Shapes

In Topic B, students combine shapes to form composite shapes, which in turn get larger as they add yet more shapes. As students work toward the objectives within the topic, they informally explore relationships between parts and wholes.

1 hexagon

Lessons 4 and 5 build on students' knowledge of attributes of shapes to create composite shapes. In Lesson 4, students create composite shapes (hexagons, rectangles, and trapezoids) from triangles, squares, and rectangles. The students recognize that the same composite shape (whole) can be made from a variety of shapes (parts). For example, a hexagon might be made by composing six triangles or two trapezoids or one trapezoid and three triangles. Students also use square tiles to see that a large rectangle can have many combinations of smaller rectangles within it.

2 trapezoids

2 triangles and 2 rhombuses

In Lesson 5, students begin by identifying the hidden shapes within a large square as they cut the seven tangram pieces from this special rectangle. Students use the pieces to form new shapes from composite shapes, including recomposing the original square. Students explore the variety of ways they can compose new shapes by positioning pieces alongside composite shapes.

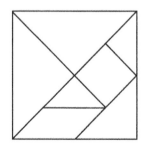

For example, students can not only form a larger triangle from two right triangles, but they can also use the shapes to form two composite triangles and push them together to make the original tangram square. Students also see how the same shapes can form different composite shapes. For instance, the same two right triangle pieces that formed a larger triangle can also be repositioned to form a square or parallelogram.

In Lesson 6, students extend their exploration of parts and wholes to three-dimensional shapes. Students create and hide composite shapes and describe the shape to a partner using attributes and positional words. The partner listens and attempts to create the same composite shape. In this way, students attend to the parts within the whole of their created shape and continue to develop clear, precise language.

A Teaching Sequence Toward Mastery of Part–Whole Relationships Within Composite Shapes
Objective 1: Create composite shapes from two-dimensional shapes.   (Lesson 4)
Objective 2: Compose a new shape from composite shapes.   (Lesson 5)
Objective 3: Create a composite shape from three-dimensional shapes and describe the composite shape using shape names and positions.   (Lesson 6)

Lesson 4

Objective: Create composite shapes from two-dimensional shapes.

Suggested Lesson Structure

- ■ Fluency Practice   (13 minutes)
- ■ Application Problem   (7 minutes)
- ■ Concept Development   (30 minutes)
- ■ Student Debrief   (10 minutes)
-   **Total Time**   **(60 minutes)**

Fluency Practice (13 minutes)

- Grade 1 Core Fluency Differentiated Practice Sets **1.OA.6**   (5 minutes)
- Number Bond Addition and Subtraction **1.OA.6**   (5 minutes)
- Shape Flash **1.G.1**   (3 minutes)

Grade 1 Core Fluency Differentiated Practice Sets (5 minutes)

Materials: (S) Core Fluency Practice Sets (Lesson 3 Core Fluency Practice Sets)

Note: Give the appropriate Practice Set to each student. Students who are repeating a set should be motivated to try to improve their performance.

Students complete as many problems as they can in 90 seconds. Assign a counting pattern and start number for early finishers, or tell them to practice make ten addition and subtraction on the back of their papers. When time runs out, collect and correct any Practice Sets that are completed.

Number Bond Addition and Subtraction (5 minutes)

Materials: (S) Personal white board, 1 die per pair

Note: This fluency activity addresses Grade 1's core fluency of sums and differences through 10 and strengthens understanding of the relationship between addition and subtraction.

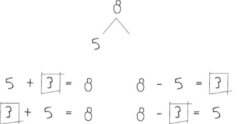

- Assign partners of equal ability and an appropriate range of numbers for each pair.
- Allow partners to choose a number for their whole greater than or equal to 6 and roll the die to determine one of the parts.

- Both students write two addition and two subtraction sentences with a box for the unknown number in each equation and solve for the missing number.
- They then exchange boards and check each other's work.

Shape Flash (3 minutes)

Materials: (T) Two-dimensional shape flash cards (Fluency Template), three-dimensional shapes used in Lesson 3

Note: This fluency activity reviews the attributes and names of two-dimensional (trapezoid, rhombus, square, rectangle, triangle) and three-dimensional (cone, cube, cylinder, sphere, rectangular prism) shapes. For three-dimensional shapes, hold up a sample of the shape, rather than a picture of the shape. As soon as students are ready to visualize, flash the shape instead.

Flash a shape card or a three-dimensional shape for three seconds. Ask a question to review an attribute or a vocabulary word students learned over the past few lessons. Pause long enough to provide thinking time, and then snap to signal students to answer.

Alternate between flashing a two-dimensional shape flash card or a three-dimensional shape. For three-dimensional shapes, ask questions such as the ones listed below:

- What's it called?
- How many faces did you see?
- How many points did this shape have?
- How many faces were square?
- Was the shape open or closed?

NOTES ON MULTIPLE MEANS OF ACTION AND EXPRESSION:

Some students may find this fluency activity challenging or need more time finishing problems. Scaffold tasks by carefully selecting the number of problems to be completed for certain partners. Give students who need a challenge a pair of dice, and have them choose a whole equal to or greater than 12 and roll the dice to find one of the parts.

Application Problem (7 minutes)

Anton made a tower 5 cubes high. Ben made a tower 7 cubes high. How much taller is Ben's tower than Anton's?

Note: If students struggled with the *compare with difference unknown* problem in Lesson 3, use a guided approach. Have students follow the steps outlined below:

- Read the story's first two sentences.
- Draw and label a picture.

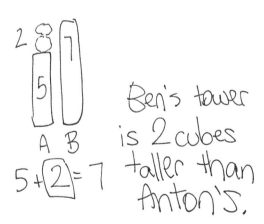

Lesson 4: Create composite shapes from two-dimensional shapes.

- Analyze their drawing. *Who has the taller tower? How many more cubes does Anton need to have a tower as tall as Ben's?*
- Read the question: *How much taller is Ben's tower than Anton's?*
- Reflect on their peers' work. Show two students' drawings and strategies using the document camera, and invite them to explain how they solved the problem.

Concept Development (30 minutes)

Materials: (T) Pattern blocks, chart paper, colored marker (S) Pattern blocks (set of 1–2 hexagons, 6 squares, 6–10 triangles, 2–4 trapezoids, 2–4 blue rhombuses, 2–4 tan rhombuses), personal white board (optional)

Note: Students use the same set of pattern blocks during Lesson 7. It may be useful for students to place a set in their personal toolkit. Tell students that pattern blocks are actually three-dimensional solids that the class will use to create two-dimensional shapes. For example, when a student traces the yellow shape, he gets a two-dimensional hexagon on the paper.

Distribute materials, and have students seated at their desks or tables.

T: For the next few days, we will be using pattern blocks to learn more about shapes. Take two minutes to explore the kinds of shapes you can make using these materials.

As students explore, walk around and take note of examples that can be used during the lesson. If available, use a camera that can be easily plugged in to display images on the board to take pictures of any compositions that might be useful, or have students compose shapes on their personal white boards to easily share with the class.

T: What shapes do the pattern blocks come in?
S: Hexagons. → Squares. → Triangles. → Trapezoids. → Two different types of rhombuses.
T: Do we have any rectangles?
S: A square is a special kind of rectangle!
T: You're right. We DO have a special kind of rectangle in the square. The square is also a special kind of rhombus, so we actually have three types of rhombuses.

1 hexagon

T: Most of you made lots of bigger shapes, or **composite shapes**, by putting the pieces together. Try to make a larger rectangle using your squares.
S: (Use squares of varying number to make a rectangle.)
T: How did you make a larger rectangle?
S: I put two squares next to each other. → I used all of my squares to make it really long.

2 trapezoids

T: I'm going to record the composite shapes that you are making. (Draw on chart paper while describing.) One person used two squares to make this size rectangle. Another person used four squares. (Quickly label the inside of each shape with the part, such as square. Label the outside of the shape with the word *rectangle*.) This whole rectangle is made of four parts that are squares. (Trace the whole rectangle with the colored marker.) Great!

T: Let's move all of our blocks to the side and take out the hexagon. (Wait.)

T: Many of you made this same hexagon shape using other pieces. Try this again. Cover the hexagon with other shapes so that you make the exact same shape with other parts. (Give students 30 seconds or more to create a hexagon shape using parts.)

T: Tell your partner what parts you used to make the hexagon.

S: I used six triangles. → I used two trapezoids. → I used three blue rhombuses.

T: Did anyone use different types of parts to make the hexagon?

S: I used two rhombuses and two triangles. → I used one trapezoid and three triangles. → I used four triangles and one rhombus.

T: Is there only one way to make one whole hexagon?

S: No!

2 triangles and 2 rhombuses

Repeat the process with other composite shapes that can be named, such as the following: a large triangle, a large rhombus, a large square, and a large trapezoid.

T: Now, move your pieces to the side again, and take out all of your square pieces. Make a rectangle with two rows of squares using all of your pieces. (Wait as students assemble the rectangle.)

T: How many small squares are in this rectangle?

S: Six squares. (Touch while counting the small squares together.)

T: Now, look closely. How many larger squares can you find hiding in the rectangle? Talk with a partner to decide. (Give students 20 to 30 seconds to look for the two larger squares.)

T: Where did you find a larger square?

S: The first four squares put together make a larger square. → If you start at the other end, you can make a square with the last four squares.

T: Great job! You can make six little squares from this rectangle, or you can make one large square using this side of the rectangle. (Point to or slide over the section on the left that forms a large square to help students visualize. Then, slide this large square back.) You can also make one large square using *this* side of the rectangle. (Point to or slide over the section on the right that forms a large square to help students visualize. Then, slide this large square back.)

NOTES ON
MULTIPLE MEANS
OF REPRESENTATION:

There are many directions to follow during this part of the lesson. Be sure to guide English language learners and students who have difficulty following multiple-step directions. These students would benefit from visual cues or possibly working with a partner at this time.

Lesson 4: Create composite shapes from two-dimensional shapes.

T: How many squares did we find all together?
S: Eight squares!
T: Although our composite shape of this rectangle is made of six squares, there are also larger squares composed of the smaller squares. Great detective work!

Problem Set (10 minutes)

Students should do their personal best to complete the Problem Set within the allotted 10 minutes. For some classes, it may be appropriate to modify the assignment by specifying which problems they work on first. Some problems do not specify a method for solving.

Student Debrief (10 minutes)

Lesson Objective: Create composite shapes from two-dimensional shapes.

The Student Debrief is intended to invite reflection and active processing of the total lesson experience.

Invite students to review their solutions for the Problem Set. They should check work by comparing answers with a partner before going over answers as a class. Look for misconceptions or misunderstandings that can be addressed in the Debrief. Guide students in a conversation to debrief the Problem Set and process the lesson.

Any combination of the questions below may be used to lead the discussion.

- Compare Problem 1 and Problem 3. What do you notice?
- Look at Problem 4. How could your shape from Problem 1 help you come up with a new way to make a hexagon using the pattern blocks?
- Look at Problem 6. How many people found at least 20 squares? 22 squares? 25 squares? Can you find more squares than you have found so far? Work with a partner to share the squares you found and see if there are more that you can find together.

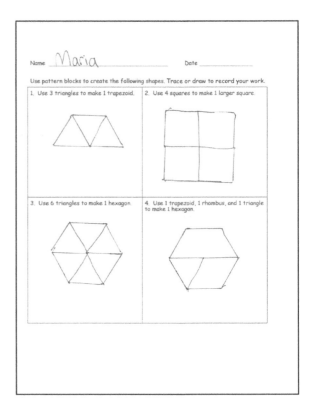

68   Lesson 4:   Create composite shapes from two-dimensional shapes.

- Look at the picture you made in Problem 7. What **composite shapes**, or larger shapes made from smaller shapes, can you name within your picture? What smaller shapes were used to make these larger shapes?
- Think about today's Fluency Practice. Name at least one addition problem that slows you down. Does anyone have a way to know that fact more easily?

Exit Ticket (3 minutes)

After the Student Debrief, instruct students to complete the Exit Ticket. A review of their work will help with assessing students' understanding of the concepts that were presented in today's lesson and planning more effectively for future lessons. The questions may be read aloud to the students.

Lesson 4: Create composite shapes from two-dimensional shapes.

Name _____   Date _____

Use pattern blocks to create the following shapes. Trace or draw to record your work.

1. Use 3 triangles to make 1 trapezoid.	2. Use 4 squares to make 1 larger square.
3. Use 6 triangles to make 1 hexagon.	4. Use 1 trapezoid, 1 rhombus, and 1 triangle to make 1 hexagon.

A STORY OF UNITS   Lesson 4 Problem Set 1•5

5. Make a rectangle using the squares from the pattern blocks. Trace the squares to show the rectangle you made.

6. How many squares do you see in this rectangle?

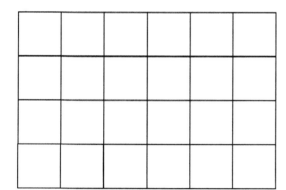

I can find _____ squares in this rectangle.

7. Use your pattern blocks to make a picture. Trace the shapes to show what you made. Tell a partner what shapes you used. Can you find any larger shapes within your picture?

Lesson 4 Exit Ticket 1•5

Name _____ Date _____

Use pattern blocks to create the following shapes. Trace or draw to show what you did.

1. Use 3 rhombuses to make a hexagon.	2. Use 1 hexagon and 3 triangles to make a large triangle.

A STORY OF UNITS   Lesson 4 Homework 1•5

Name _____   Date _____

Cut out the pattern block shapes from the bottom of the page. Color them to match the key, which is different from the pattern block colors in class. Trace or draw to show what you did.

| Hexagon—red   Triangle—blue   Rhombus—yellow   Trapezoid—green |

| 1. Use 3 triangles to make 1 trapezoid. | 2. Use 3 triangles to make 1 trapezoid, and then add 1 trapezoid to make 1 hexagon. |

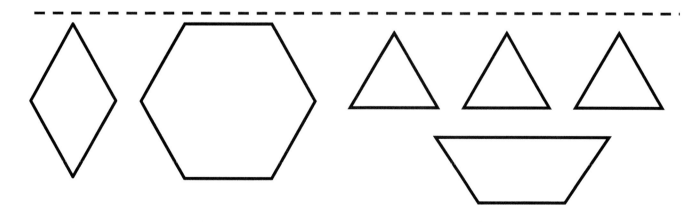

Lesson 4:   Create composite shapes from two-dimensional shapes.

3. How many squares do you see in this large square?

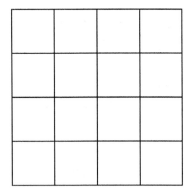

I can find _____ squares in this rectangle.

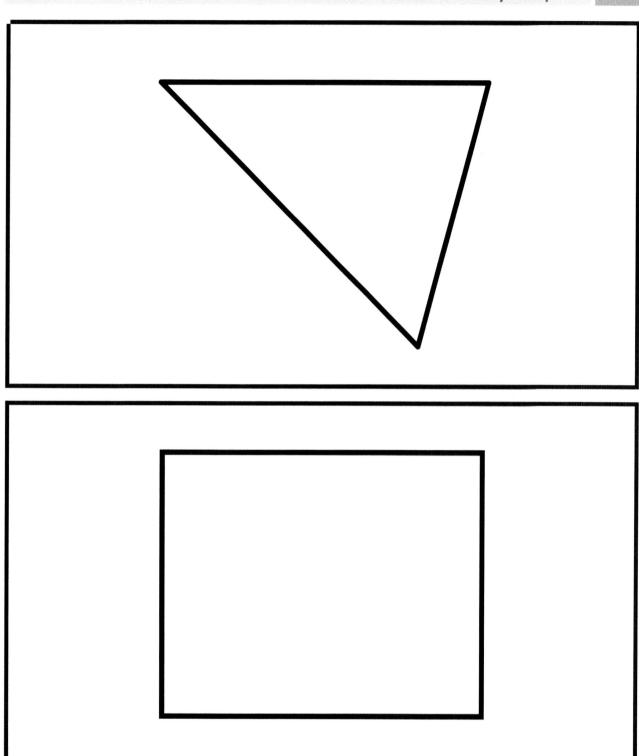

two-dimensional shape flash cards

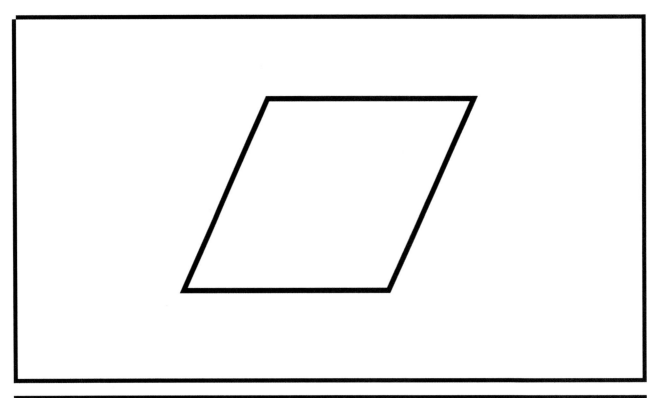

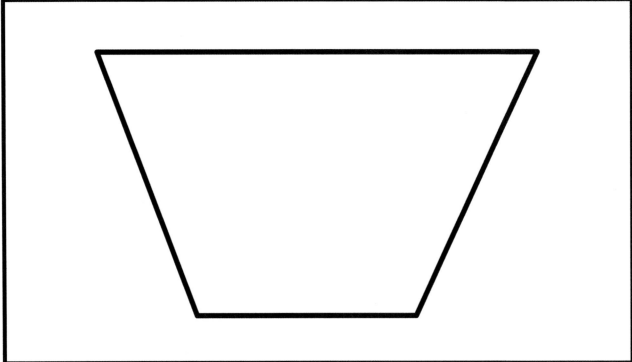

two-dimensional shape flash cards

Lesson 4 Fluency Template

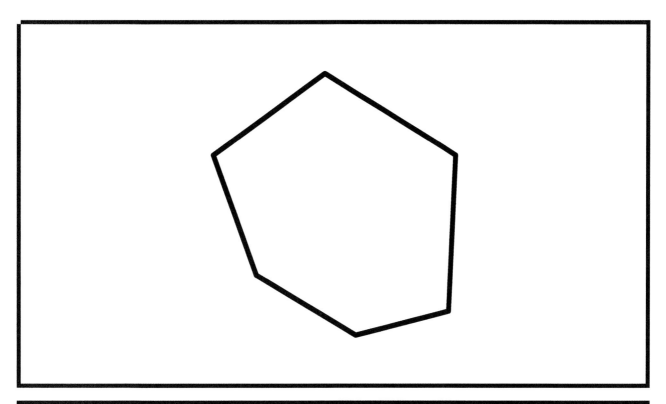

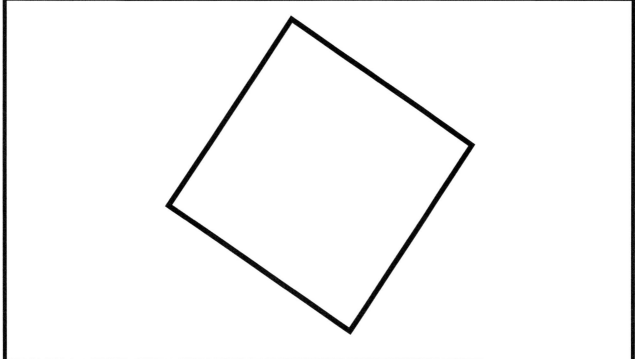

two-dimensional shape flash cards

Lesson 4: Create composite shapes from two-dimensional shapes.

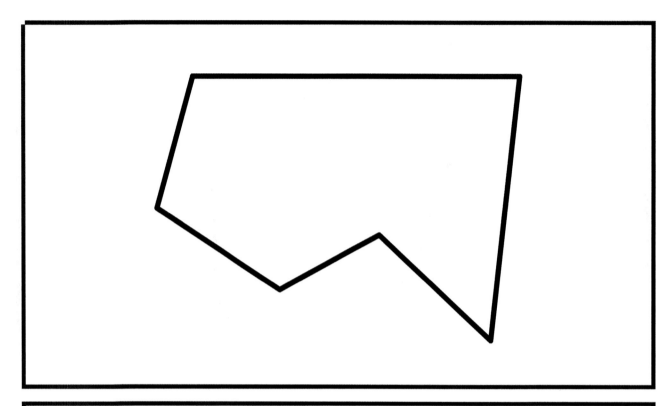

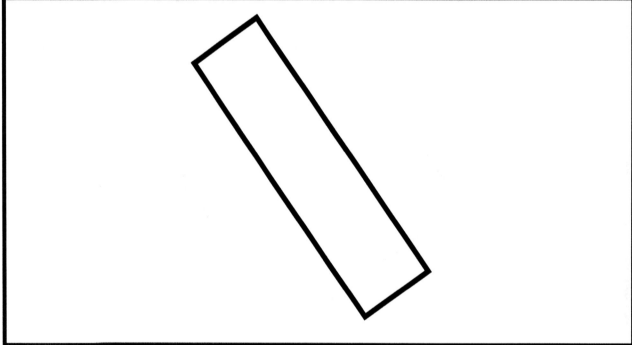

two-dimensional shape flash cards

Lesson 5

Objective: Compose a new shape from composite shapes.

Suggested Lesson Structure

- ■ Fluency Practice   (13 minutes)
- ▨ Application Problem (5 minutes)
- ▢ Concept Development (32 minutes)
- ■ Student Debrief   (10 minutes)
- **Total Time**   **(60 minutes)**

A NOTE ON STANDARDS ALIGNMENT:

In this lesson, students use tangram pieces as a context for composing a new shape from composite shapes (**1.G.2**). The Progression Document on Geometry does not include parallelogram as a shape for Grade 1 students, although this shape is one of the basic shapes within a tangram.

Fluency Practice (13 minutes)

- Grade 1 Core Fluency Sprint **1.OA.6**   (10 minutes)
- Shape Flash **1.G.1**   (3 minutes)

Grade 1 Core Fluency Sprint (10 minutes)

Materials: (S) Core Fluency Sprint (Lesson 1 Core Fluency Sprint)

Note: Choose an appropriate Sprint, based on the needs of the class. Motivate students to monitor and appreciate their own progress. As students work, observe the areas where they slow down or get stuck. Pay attention to the strategies students use.

Core Fluency Sprint List:

- Core Addition Sprint 1 (Targets core addition and missing addends.)
- Core Addition Sprint 2 (Targets the most challenging addition within 10.)
- Core Subtraction Sprint (Targets core subtraction.)
- Core Fluency Sprint: Totals of 5, 6, and 7 (Develops understanding of the relationship between addition and subtraction.)
- Core Fluency Sprint: Totals of 8, 9, and 10 (Develops understanding of the relationship between addition and subtraction.)

Shape Flash (3 minutes)

Materials: (T) Two-dimensional shape flash cards (Lesson 4 Fluency Template), three-dimensional shapes used in Lesson 3

Note: This fluency activity reviews the attributes and names of two-dimensional and three-dimensional shapes. For three-dimensional shapes, consider displaying the shape as students answer the questions. As soon as students are ready to visualize, flash the shape instead. Repeat Shape Flash from Lesson 4.

Lesson 5:   Compose a new shape from composite shapes.

Lesson 5

Application Problem (5 minutes)

Darnell and Tamra are comparing their grapes. Darnell's vine has 9 grapes. Tamra's vine has 6 grapes. How many more grapes does Darnell have than Tamra?

Note: This *compare with difference unknown* problem continues to engage students in the same type of problem using different contexts and a larger difference between the numbers. If necessary, remind students that they are comparing Darnell's grapes and Tamra's grapes. When comparing two numbers, it is best to use double tape diagrams, which more clearly support visualizing the difference between the two quantities.

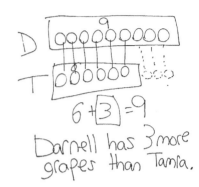

Concept Development (32 minutes)

Materials: (T) Tangram (Template), scissors (S) Tangram (Template) (cut off the bottom tangram on each sheet to be sent home with homework), scissors

Note: This lesson uses tangrams. If time allows, consider sharing the origin of the tangram or read *Grandfather Tang's Story* by Ann Tompert. Of the 7 individual pieces within a tangram, there is one shape that is not part of the Grade 1 Standards for Geometry: parallelogram. For this reason, the attributes of a parallelogram are not discussed in this lesson. Students are introduced to the shape name as a way to discuss the pieces being used as they create composite shapes.

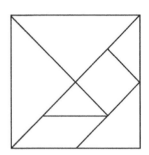

Have students store their tangram pieces in their personal toolkits to be used during Lesson 7.

Cut out the large square from the tangram sheet to be used by the teacher. Distribute the materials to students, seated at their desks or tables.

NOTES ON MULTIPLE MEANS OF ENGAGEMENT:

Some students may need support cutting their tangram sheets. Precut some of the sheets, or, as the rest of the class is cutting, assist certain students.

T: Today, we will be cutting out our shapes from this one large shape. What is this shape? (Hold up the tangram backward, so students do not see all of the lines within the square.)

S: A square.

T: Cut out the large square from your piece of paper. (Wait as students cut.)

T: Look how I folded my paper down the diagonal line that goes through the middle of the square. (Fold paper.) What do you see on one side?

S: A triangle!

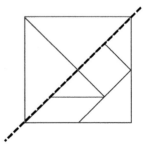

Lesson 5: Compose a new shape from composite shapes.

A STORY OF UNITS   Lesson 5   1•5

T: Cut out this triangle on your paper as I cut out my triangle. (Cut out the triangle as students cut out the triangle.)
T: How many pieces do you have now?
S: Two pieces!
T: What is the shape of each piece?
S: They are both triangles!
T: Both of these triangles are made of smaller parts. What parts do you see in this triangle? (Hold up the triangle made of two triangles.)
S: That triangle is made of two smaller triangles.
T: What parts do you see in this triangle? (Hold up the other triangle.)
S: I see two small triangles and one bigger triangle. → I see a square. → I see another shape. It kind of looks like a rhombus, but the sides don't look like they are the same length.
T: You are right. That shape (point to the parallelogram) is not quite a rhombus. A rhombus is a special parallelogram that has equal straight sides. When the shape is like this, where all pairs of opposite sides are equal, it is called a parallelogram. Do you see how this pair is not the same length as this pair? One pair is long, and the other is shorter, so it cannot be called a rhombus. We just call it a parallelogram.
T: Let's cut apart the two triangles that make this first large triangle. (Fold the larger triangle in half to show the two smaller triangles, and cut. Students do the same.)
T: Put your two triangles you cut apart to the side. Take the largest triangle on your table, and place it in front of you like mine. (Place the longest side as the base.) Let's cut off this little triangle at the top. (Students and teacher all cut off the top triangle.)
T: What shape do the square, little triangles, and parallelogram make together? Do you see what shape has been hiding inside the larger triangle?
S: A trapezoid!
T: Let's cut out the parallelogram. (Cut and circulate as students are cutting.)
T: What shape do the two triangles and the square make together?
S: A smaller trapezoid!
T: Now, let's cut apart all of the last pieces. (Cut and circulate as students are cutting.)
T: Put your pieces back to form the large square we started with. (Allow students ample time to position the pieces. Make every effort not to interfere as students work at positioning the shapes during this sequence of the lesson. Encourage students to persevere, providing the least direction possible. For students who finish quickly, have them shuffle their pieces and try to make new shapes.)

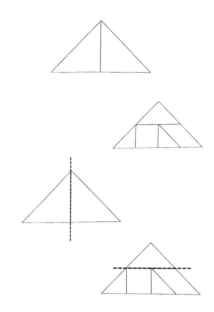

NOTES ON
MULTIPLE MEANS
OF ENGAGEMENT:

Arranging shapes to create specific composite shapes and then recomposing the pieces into different shapes can be challenging for some students. Be sure to allow time for students to use models or work with a partner as these students work on coordinating their visual and motor skills.

   Lesson 5:   Compose a new shape from composite shapes.   81

A STORY OF UNITS

Lesson 5 1•5

T: Great job! These seven pieces that form a large square are called a tangram. You can make lots of different and interesting shapes by combining some or all of the parts. Let's use just the two largest triangles. Put all the other pieces to the side. (Wait as students move pieces.)

T: If I leave these pieces the way they were, what shape do they make when they are together?

S: A large triangle!

T: Move the shapes around, and see if you can make another shape using the same pieces. (Circulate as students work individually or with a partner.)

T: What shape did you make?

S: I made a square. → I made a parallelogram.

T: With your partner, take two or three of the same tangram pieces, and try to each make a different shape using the same pieces. Here's a hint: You may want to flip over your pieces, turn them, or slide them around to make the new shapes.

After students have worked with their partners for two or three minutes, have pairs share one of the various composite shapes they made.

Problem Set (10 minutes)

Students should do their personal best to complete the Problem Set within the allotted 10 minutes. For some classes, it may be appropriate to modify the assignment by specifying which problems they work on first. Some problems do not specify a method for solving.

Student Debrief (10 minutes)

Lesson Objective: Compose a new shape from composite shapes.

The Student Debrief is intended to invite reflection and active processing of the total lesson experience.

Invite students to review their solutions for the Problem Set. They should check work by comparing answers with a partner before going over answers as a class. Look for misconceptions or misunderstandings that can be addressed in the Debrief. Guide students in a conversation to debrief the Problem Set and process the lesson.

Any combination of the questions below may be used to lead the discussion.

- Which shapes are used to make the large square we call a tangram? Which smaller shapes can be seen inside the tangram square?
- Look at Problem 2. Share how you made a square. Could you have used other tangram pieces to make the square?

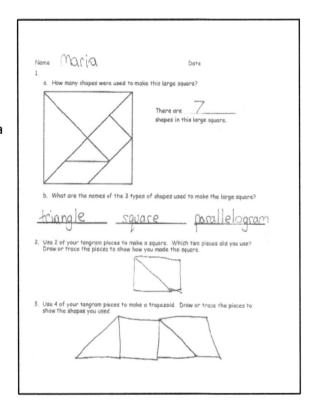

82   Lesson 5:   Compose a new shape from composite shapes.

EUREKA MATH

- Look at Problem 3. Share how you made a trapezoid with four pieces. Could you have made a trapezoid with fewer pieces? Demonstrate your solution. Compare the similarities and differences.
- How did you cover the picture in Problem 4? Did everyone use the same pieces in the same places? Why or why not?
- Think about today's Fluency Practice. Did you get better at a *slow-me-down* problem today? Did you do anything to make that happen?

Exit Ticket (3 minutes)

After the Student Debrief, instruct students to complete the Exit Ticket. A review of their work will help with assessing students' understanding of the concepts that were presented in today's lesson and planning more effectively for future lessons. The questions may be read aloud to the students.

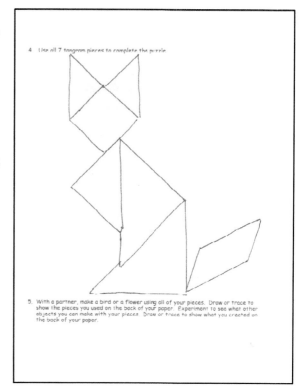

Lesson 5: Compose a new shape from composite shapes.

Name _____ Date _____

1.
   a. How many shapes were used to make this large square?

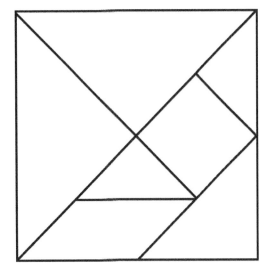

   There are _____ shapes in this large square.

   b. What are the names of the 3 types of shapes used to make the large square?

   _____   _____   _____

2. Use 2 of your tangram pieces to make a square. Which 2 pieces did you use? Draw or trace the pieces to show how you made the square.

3. Use 4 of your tangram pieces to make a trapezoid. Draw or trace the pieces to show the shapes you used.

4. Use all 7 tangram pieces to complete the puzzle.

5. With a partner, make a bird or a flower using all of your pieces. Draw or trace to show the pieces you used on the back of your paper. Experiment to see what other objects you can make with your pieces. Draw or trace to show what you created on the back of your paper.

Name _____ Date _____

Use words or drawings to show how you can make a larger shape with 3 smaller shapes. Remember to use the names of the shapes in your example.

A STORY OF UNITS   Lesson 5 Homework 1•5

Name _____ Date _____

1. Cut out all of the tangram pieces from the separate piece of paper you brought home from school. It looks like this:

2. Tell a family member the name of each shape.

3. Follow the directions to make each shape below. Draw or trace to show the parts you used to make the shape.

   a. Use 2 tangram pieces to make 1 triangle.

   b. Use 1 square and 1 triangle to make 1 trapezoid.

   c. Use one more piece to change the trapezoid into a rectangle.

Lesson 5:   Compose a new shape from composite shapes.

4. Make an animal with all of your pieces. Draw or trace to show the pieces you used. Label your drawing with the animal's name.

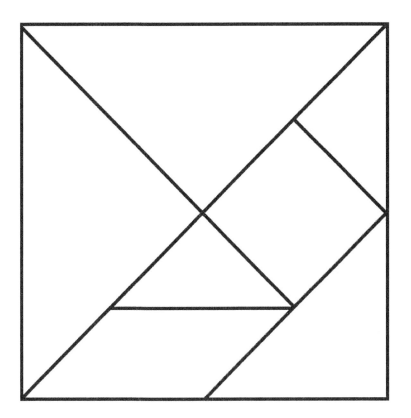

One tangram is to be used during class.
The other tangram is to be sent home with the homework.

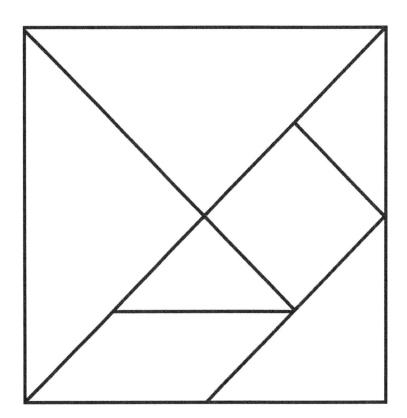

tangram

Lesson 5: Compose a new shape from composite shapes.

Lesson 6

Objective: Create a composite shape from three-dimensional shapes and describe the composite shape using shape names and positions.

Suggested Lesson Structure

■ Fluency Practice   (13 minutes)
■ Application Problem   (5 minutes)
■ Concept Development   (32 minutes)
■ Student Debrief   (10 minutes)
  Total Time   **(60 minutes)**

Fluency Practice (13 minutes)

- Grade 1 Core Fluency Sprint **1.OA.6**   (10 minutes)
- Coin Drop **1.OA.6, 1.NBT.6**   (3 minutes)

Grade 1 Core Fluency Sprint (10 minutes)

Materials: (S) Core Fluency Sprint (Lesson 1 Core Fluency Sprint)

Note: Based on the needs of the class, select a Core Fluency Sprint. Consider the options below:

- Re-administer the previous lesson's Sprint.
- Administer the next Sprint in the sequence.
- Differentiate. Administer two different Sprints. Simply have one group do a counting activity on the back of the first Sprint as the other group corrects the second Sprint.

NOTES ON MULTIPLE MEANS OF ENGAGEMENT:

Encourage students to set goals for improvement on Sprints and Fluency Practice Sets. Provide scaffolds, strategies, and opportunities for practice to help students reach their personal goals.

Coin Drop (3 minutes)

Materials: (T) 4 dimes, 10 pennies, can

Note: In this activity, students practice adding and subtracting ones and tens.

T: (Hold up a penny.) Name my coin.
S: A penny.

NOTES ON MULTIPLE MEANS OF ENGAGEMENT:

After playing Coin Drop with pennies and then dimes, mix pennies and dimes so that students have to add based on the changing value of the coin. This challenges students and keeps them listening for what comes next.

A STORY OF UNITS   Lesson 6 1•5

T: How much is it worth?
S: 1 cent.
T: Listen carefully as I drop pennies in my can. Count along in your minds.

Drop in some pennies, and ask how much money is in the can. Take out some pennies, and show the class. Ask how much money is still in the can. Continue adding and subtracting pennies for a minute or so. Then, repeat the activity with dimes.

Application Problem (5 minutes)

Emi lined up 4 yellow cubes in a row. Fran lined up 7 blue cubes in a row. Who has fewer cubes? How many fewer cubes does she have?

Note: Today's Application Problem continues to provide the opportunity for students to work with *compare with difference unknown* problem types. For the past few days, students have looked at questions that asked *how many more*. Today's question incorporates the challenging vocabulary word *fewer*. Consider giving examples of the word *fewer* prior to having students solve the problem.

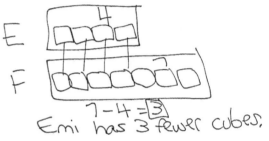

Concept Development (32 minutes)

Materials: (T) Three-dimensional solids including cubes, cones, rectangular prisms, spheres, and cylinders; 1 large privacy folder (S) Sets of three-dimensional shapes, large privacy folder (1 per pair)

Note: This lesson works best with ample materials for each set of students. If a set of three-dimensional solids is not readily available, use a collection of reused or recycled materials such as those listed in Lesson 3.

On a table or desk, behind a privacy folder, gather the teacher's set of three-dimensional shapes so that students cannot see the shapes as the teacher picks them up to build. Distribute the materials to students, seated at their desks or tables. Place one additional sample of each shape on the floor or table in front of the class for students who need visual reminders of each shape.

T: I am going to build a three-dimensional structure but hide it behind this folder. Listen to my description, and try to build the same shape at your desk.

T: (Slowly describe the structure, providing time for students to build while the teacher explains each shape's placement.)

NOTES ON
MULTIPLE MEANS
OF ENGAGEMENT:

While describing the composite structure students are to build, consider giving visual cues for certain words. Otherwise, some students may not be able to keep up with the directions. Some will benefit from directional cues or seeing the shape they are supposed to place on their desks.

Lesson 6:   Create a composite shape from three-dimensional shapes and describe the composite shape using shape names and positions.

91

T: I am putting…
- A cube on the table.
- A cone on top of the cube so that the circular face is touching the top of the cube.

T: Do you think your structure looks like my structure? Share what you built with your partner.

S: (Discuss choices.)

T: (Remove the privacy folder to reveal the structure.) Were you correct?

S: Yes! (Allow students a moment to adjust their structures if they were not correct.)

Repeat the process, building a structure with three components, such as the following:

T: I am putting…
- A rectangular prism with the longest face touching the table.
- A cube on top of the prism, directly in the middle.
- Two cones on top of the prism, one on each end.

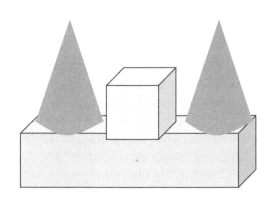

Repeat the process a third time, as described below:

T: (Slowly describe the structure, providing time for students to build while explaining each shape's placement.)

T: I am putting…
- A rectangular prism with the longest face touching the table.
- A cylinder on top of the prism all the way to the right, with the circular face touching the prism.
- A cylinder on top of the prism all the way to the left, with the circular face touching the prism.
- A rectangular prism on top of these cylinders so that it touches both cylinders.
- A cone right in the center of this rectangular prism, with the circular face touching the prism.

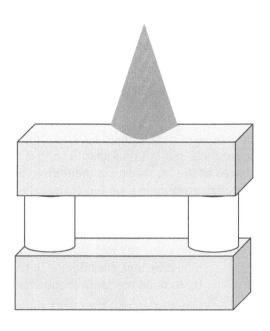

T: Let me repeat my description. As I do, look at your structure, and decide if you have everything where you want it. (Repeat the description as students check the structures they have created.)

T: Who is convinced they have the same structure that I have? Explain why you think you are correct.

S: (Use reasoning, along with the description that was provided.)

T: (Remove the privacy folder to reveal the structure.) Were you correct?
S: Yes! (Allow students a moment to adjust their structures if they were not correct.)
T: Do you like my new composite shape?
S: Yes!
T: Do you think you could make and describe your own interesting composite shapes?
S: Yes!
T: With your partner, you are going to get to play the Build My Composite Shape Game that we just played together.

- Partner A will make a structure behind his hiding folder. Partner B should turn her back so that he cannot peek. Partner A will tell Partner B when to turn around.
- As Partner A describes the structure, Partner B tries to make it with her three-dimensional shapes. When she thinks she has the right structure, Partner A removes the folder, and they compare structures.
- The partners switch roles. Continue to take turns until time is up.

As students play, circulate and ensure that students are using precise language to describe the position and location of their three-dimensional shapes. When partners are building different structures, ask Partner A to explain the location and position of the shapes again, and support clear communication between students.

Problem Set (10 minutes)

Students should do their personal best to complete the Problem Set within the allotted 10 minutes. For some classes, it may be appropriate to modify the assignment by specifying which problems they work on first.

Student Debrief (10 minutes)

Lesson Objective: Create a composite shape from three-dimensional shapes and describe the composite shape using shape names and positions.

The Student Debrief is intended to invite reflection and active processing of the total lesson experience.

Invite students to review their solutions for the Problem Set. They should check work by comparing answers with a partner before going over answers as a class. Look for misconceptions or misunderstandings that can be addressed in the Debrief. Guide students in a conversation to debrief the Problem Set and process the lesson.

Any combination of the questions below may be used to lead the discussion.

- Which three-dimensional shapes did you choose to use and why? Why did you choose to leave some shapes out?
- Were more spheres or cubes used in the structures? Why might that be?
- Find two three-dimensional structures that used the same pieces to make different larger shapes. Explain the similarities and differences.
- Look at today's Sprint. Explain how the answer to the first number sentence helped you easily solve the next number sentence.

Exit Ticket (3 minutes)

After the Student Debrief, instruct students to complete the Exit Ticket. A review of their work will help with assessing students' understanding of the concepts that were presented in today's lesson and planning more effectively for future lessons. The questions may be read aloud to the students.

Lesson 6 Problem Set 1•5

Name _____   Date _____

1. Work with your partner and another pair to build a structure with your 3-dimensional shapes. You can use as many of the pieces as you choose.

2. Complete the chart to record the number of each shape you used to make your structure.

Cubes	
Spheres	
Rectangular Prisms	
Cylinders	
Cones	

3. Which shape did you use on the bottom of your structure? Why?

4. Is there a shape you chose not to use? Why or why not?

Name _____   Date _____

Maria made a structure using her 3-dimensional shapes. Use your shapes to try to make the same structure as Maria as your teacher reads the description of Maria's structure.

Maria's structure has the following:

- 1 rectangular prism with the shortest face touching the table.
- 1 cube on top and to the right of the rectangular prism.
- 1 cylinder on top of the cube with the circular face touching the cube.

A STORY OF UNITS   **Lesson 6 Homework**   **1•5**

Name _____ Date _____

Use some 3-dimensional shapes to make another structure. The chart below gives you some ideas of objects you could find at home. You can use objects from the chart or other objects you may have at home.

Cube	Rectangular prism	Cylinder	Sphere	Cone
Block	Food box: Cereal, macaroni and cheese, spaghetti, cake mix, juice box	Food can: Soup, vegetables, tuna fish, peanut butter	Balls: Tennis ball, rubber band ball, basketball, soccer ball	Ice cream cone
Dice	Tissue box	Toilet paper or paper towel roll	Fruit: Orange, grapefruit, melon, plum, nectarine	Party hat
	Hardcover book	Glue stick	Marbles	Funnel
	DVD or video game box			

Ask someone at home to take a picture of your structure. If you are unable to take a picture, try to sketch your structure or write the directions on how to build your structure on the back of the paper.

Lesson 6:   Create a composite shape from three-dimensional shapes and describe the composite shape using shape names and positions.

Mathematics Curriculum

GRADE 1 • MODULE 5

Topic C
Halves and Quarters of Rectangles and Circles

1.G.3

Focus Standard:	1.G.3	Partition circles and rectangles into two and four equal shares, describe the shares using the words *halves, fourths,* and *quarters,* and use the phrases *half of, fourth of,* and *quarter of.* Describe the whole as two of, or four of the shares. Understand for these examples that decomposing into more equal shares creates smaller shares.
Instructional Days:	3	
Coherence -Links from:	GK–M2	Two-Dimensional and Three-Dimensional Shapes
-Links to:	G2–M8	Time, Shapes, and Fractions as Equal Parts of Shapes

During Topic C, students build on their concrete work with composite shapes and begin naming equal parts of wholes, specifically halves and fourths (or quarters). Students more closely analyze the same composite shapes created in Topic B, recognizing composite shapes made from equal, non-overlapping parts and identifying halves and quarters within rectangular and circular shapes.

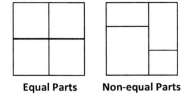

Equal Parts   Non-equal Parts

In Lesson 7, students explore composite shapes that have been made throughout the module and sort them into two categories of shapes, those made from equal parts and those made from non-equal parts. Students count the number of equal parts that form one whole.

Lesson 8 introduces the terms *half* and *quarter*, or *fourths,* to name two equal parts of a whole and four equal parts of a whole, respectively. Students learn *half-circle* and *quarter-circle* as the names of shapes and recognize that they are named for their size and shape in relation to a whole circle. Models of rectangular and circular pizzas are used for students to discuss equal parts of the whole.

In Lesson 9, students explore halves and fourths more deeply as they identify these parts within circles and rectangles of varying size and dimension. Students recognize that as they partition, or decompose, the whole into more equal shares, they create smaller units.

Topic C

A Teaching Sequence Toward Mastery of Halves and Quarters of Rectangles and Circles

Objective 1: Name and count shapes as parts of a whole, recognizing relative sizes of the parts.
(Lesson 7)

Objective 2: Partition shapes and identify halves and quarters of circles and rectangles.
(Lessons 8–9)

Lesson 7

Objective: Name and count shapes as parts of a whole, recognizing relative sizes of the parts.

Suggested Lesson Structure

- Fluency Practice   (12 minutes)
- Application Problem (5 minutes)
- Concept Development (33 minutes)
- Student Debrief   (10 minutes)

Total Time   **(60 minutes)**

Fluency Practice (12 minutes)

- Core Fluency Differentiated Practice Sets 1.OA.6   (5 minutes)
- Whisper Count 1.NBT.4   (2 minutes)
- Make Ten Addition with Partners 1.OA.6   (5 minutes)

Core Fluency Differentiated Practice Sets (5 minutes)

Materials: (S) Core Fluency Practice Sets (Lesson 3 Core Fluency Practice Sets)

Note: Give the appropriate Practice Set to each student. Students who completed all of the questions correctly on their most recent Practice Set should be given the next level of difficulty. All other students should try to improve their scores on their current levels.

Students complete as many problems as they can in 90 seconds. Assign a counting pattern and start number for early finishers, or tell them to practice make ten addition or subtraction on the back of their papers. Collect and correct any Practice Sets completed within the allotted time.

Whisper Count (2 minutes)

Materials: (T) Chart of numbers to 30 with multiples of 5 circled

Note: This activity prepares students for Lesson 11, where they add 5 minutes until reaching 30 minutes to connect half past the hour to 30 minutes past the hour. If students are proficient at counting on by fives, consider substituting for the Fluency Practice 5 More (Lesson 8).

Whisper count to 30 with students, saying multiples of 5 out loud.

A STORY OF UNITS   Lesson 7 1•5

T: Whisper count with me. Say the circled numbers out loud.
T/S: (Whisper.) 1, 2, 3, 4.
T/S: (Say.) 5.

Continue counting to 30.

Make Ten Addition with Partners (5 minutes)

Materials: (S) Personal white board

Note: This fluency activity reviews how to use the Level 3 strategy of making ten to add two single-digit numbers.

- Assign partners of equal ability.
- Partners choose an addend for each other from 1 to 10.
- On their personal white boards, students add their number to 9, 8, and 7. Remind students to write the two addition sentences they learned in Module 2, as seen in the examples below.
- Partners then exchange personal white boards and check each other's work.

NOTES ON
MULTIPLE MEANS
OF ACTION AND
EXPRESSION:

Some students may find this fluency activity challenging or need more time finishing problems. Scaffold tasks by carefully selecting the number of problems to be completed for certain partners.

Application Problem (5 minutes)

Peter set up 5 rectangular prisms to make 5 towers. He put a cone on top of 3 of the towers. How many more cones does Peter need to have a cone on every tower?

Note: This Application Problem presents a *compare with difference unknown* problem type using easy numbers. Before moving to Concept Development, link the Application Problem question with the more challenging comparison question of *How many fewer cones does Peter have than rectangular prisms?* In the student sample selected, notice that the student does not yet independently use double-tape diagrams. After the student explains how she solved this problem using her drawing, one rectangle can be drawn around the cones, and one rectangle can be drawn around the prisms, turning the drawing into a double-tape diagram. If there are students in the class who are already effectively using the double-tape diagram, the two models can be compared.

Lesson 7: Name and count shapes as parts of a whole, recognizing relative sizes of the parts.

101

Concept Development (33 minutes)

Materials: (T) Tangram pieces (Lesson 5 Template), document camera, pattern blocks, chart paper, yellow marker (S) Tangram pieces (Lesson 5 Template), pattern blocks in individual plastic bags (set of 1–2 hexagons, 6 squares, 6–10 triangles, 2–4 trapezoids, 2–4 blue rhombuses, 2–4 tan rhombuses)

Seat students at their desks or tables with the tangram pieces ready to use and pattern blocks in individual plastic bags ready for later in the lesson.

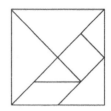

- T: Two lessons ago, we made many different shapes using two or more of these tangram pieces. Can you think of any shapes we made?
- S: We made a big square. → We made a smaller square. → We made a rectangle. → We made trapezoids and parallelograms.
- T: Great! Use two or more of your pieces to make a shape you can name.
- S: (Spend one minute creating shapes.)
- T: (Circulate and ask questions such as the following: What is the name of your overall shape? Can you add another piece to your shape to make another larger shape that you can name?)
- T: Let's look at some of the shapes you created and see what parts, or shapes, they are made of. (Choose a student who created a square using two smaller triangles. Invite the student to place his shape under the document camera.)
- T: What is the shape that he created?
- S: A square!
- T: What are the parts that he used to make this square, and how many parts are there?
- S: He used two triangles to make the square.
- T: Great! Let's record this. (Draw the shape on chart paper, partitioned to show the pieces used.) Student A used two triangles to make a square.
- T: I saw someone make a square in a different way. (Under the document camera, position all tangram pieces to make the large square.) What are the parts that are used to make this square, and how many parts are there?
- S: There are seven parts. → There are two large triangles, one medium triangle, two small triangles, one parallelogram, and one square. (Add the shapes to the chart as shown.)

NOTES ON MULTIPLE MEANS OF ACTION AND EXPRESSION:

Asking questions for comprehension during this lesson is important for guiding students toward evaluating their thinking. This provides students with an opportunity to evaluate their process and analyze errors.

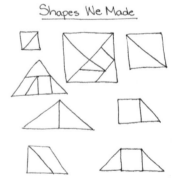

Repeat this process with any other composite shapes that students created. Some additional examples are shown in the chart.

T: Some of the shapes on our chart are made with equal parts, where two same-size parts were used to make the larger shape. Can you find them on the chart?

S: The first square is made of equal parts!

T: (Color both equal parts with a yellow marker so that the equal parts stand out.) Can you find any more shapes made with equal parts?

S: The triangle made with two smaller triangles has equal parts! (Continue as appropriate.)

T: What about the large square that we made using all of the pieces? Is this made of seven equal parts?

S: No. The parts are different sizes. There are big triangles and little triangles.

T: You are correct! Let's check the rest of our shapes on the chart to make sure we found all the shapes with equal parts. (Repeat the process by having students explain why the rest of the shapes do not have equal parts.)

T: Let's look at some of the hexagon shapes we made a few days ago. (Place one yellow hexagon pattern block under the document camera.) How can we make a hexagon using smaller pattern block pieces?

S: Use six triangles! (Place six green triangles on top of the yellow hexagon, under the document camera.)

T: Is the hexagon made of equal parts?

S: Yes!

T: How many equal parts?

S: Six!

T: What's another way to make a hexagon?

S: Two trapezoids! (Place two trapezoids on top of the yellow hexagon, under the document camera.)

T: Is the hexagon made of equal parts?

S: Yes!

T: How many equal parts?

S: Two!

T: Can we use trapezoids and triangles to make a hexagon?

S: Use one trapezoid and three triangles. (Place the pieces on top of the yellow hexagon, under the document camera.)

T: How many parts are used for this hexagon?

S: Four parts!

T: Are they four equal parts?

S: No. The trapezoid is much bigger than the triangles.

T: With a partner, make one hexagon that is created with equal parts and another hexagon that is made with parts that are not equal.

equal parts

equal parts   <u>not</u> equal parts

Lesson 7: Name and count shapes as parts of a whole, recognizing relative sizes of the parts.

Give students one minute to create composite shapes. Then, have students share their composite hexagon with the class, noting how many parts are used to make the shape and if the shape is made of equal parts. Record these shapes on the chart, coloring the composite shapes made with equal parts in yellow and labeling *2 equal parts* or *3 equal parts* as appropriate.

Extension: If time allows, invite students to use their pattern blocks to create other shapes with equal parts. The composite shapes created do not need to be shape names that students know. If including this portion, during the Student Debrief, ask students what shapes they made with their blocks and what they noticed when they used equal parts for the entire shape.

> **NOTES ON MULTIPLE MEANS OF ENGAGEMENT:**
>
> While teaching, be sure to provide cross-curricular connections for students. Visit the school or local library to check out books on shapes or equal parts to supplement learning during Topic C.

Problem Set (10 minutes)

Students should do their personal best to complete the Problem Set within the allotted 10 minutes. For some classes, it may be appropriate to modify the assignment by specifying which problems they work on first.

Student Debrief (10 minutes)

Lesson Objective: Name and count shapes as parts of a whole, recognizing relative sizes of the parts.

The Student Debrief is intended to invite reflection and active processing of the total lesson experience.

Invite students to review their solutions for the Problem Set. They should check work by comparing answers with a partner before going over answers as a class. Look for misconceptions or misunderstandings that can be addressed in the Debrief. Guide students in a conversation to debrief the Problem Set and process the lesson.

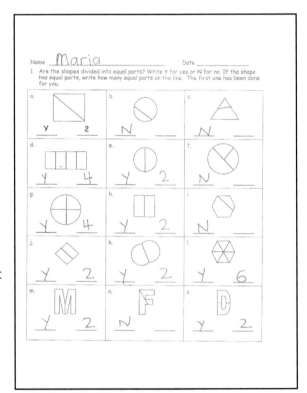

Any combination of the questions below may be used to lead the discussion.

- Look at Problem 1. Find an example of a shape that is not divided into equal parts. How did you decide that the parts were not equal?
- Look at Problem 4. What are the shapes of your equal parts? Compare with your partner. Did everyone make the same shape?

- What does it mean when we say a shape has equal parts? How is this *the same as* or *different from* the ways we have used the word *equal* in the past? Give examples of ways we use the word *equal* in math class.
- Think about your Fluency Practice today. Which addition or subtraction facts are becoming easier for you to remember?

Exit Ticket (3 minutes)

After the Student Debrief, instruct students to complete the Exit Ticket. A review of their work will help with assessing students' understanding of the concepts that were presented in today's lesson and planning more effectively for future lessons. The questions may be read aloud to the students.

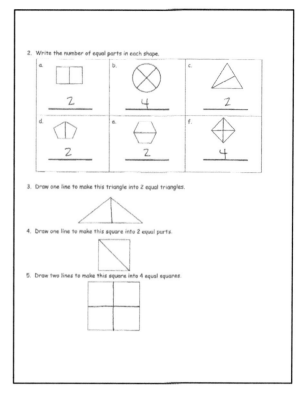

Lesson 7: Name and count shapes as parts of a whole, recognizing relative sizes of the parts.

A STORY OF UNITS

Lesson 7 Problem Set 1•5

Name _____ Date _____

1. Are the shapes divided into equal parts? Write **Y** for yes or **N** for no. If the shape has equal parts, write how many equal parts on the line. The first one has been done for you.

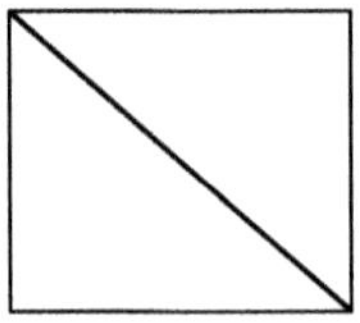 a.	b. 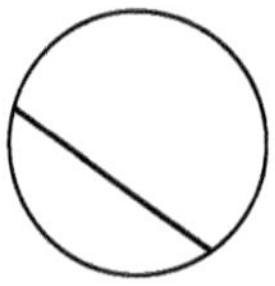	c. 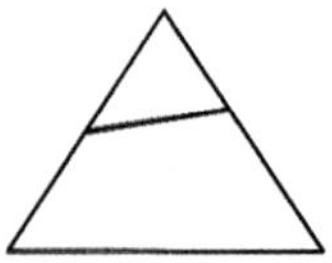
Y 2	___ ___	___ ___
d. 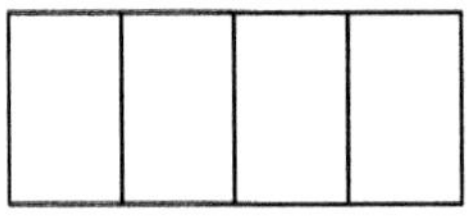	e. 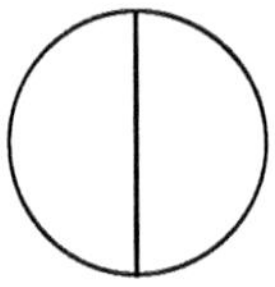	f. 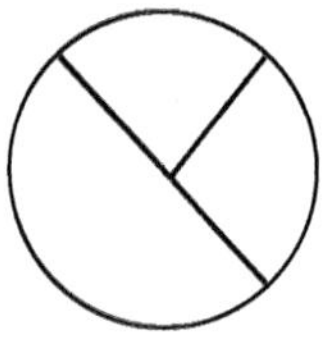
___ ___	___ ___	___ ___
g. 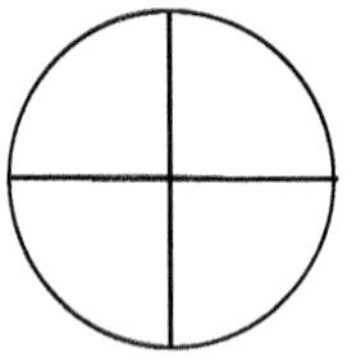	h. 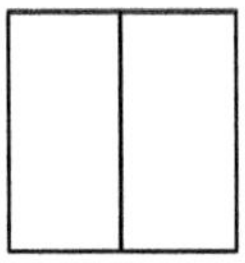	i. 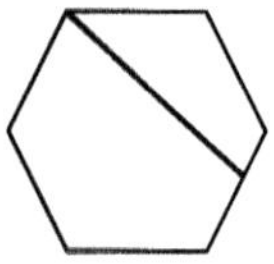
___ ___	___ ___	___ ___
j. 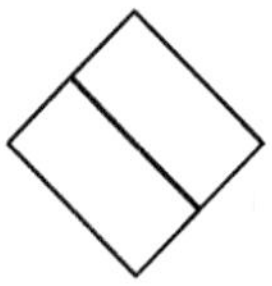	k. 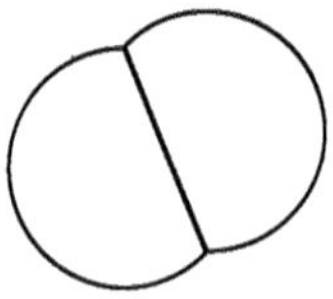	l. 
___ ___	___ ___	___ ___
m. 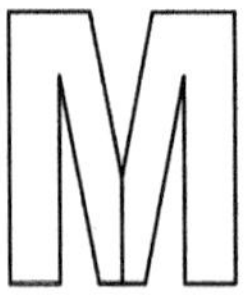 M	n. 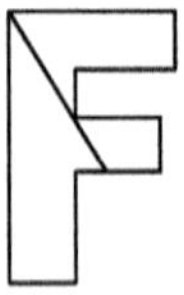 F	o.  D
___ ___	___ ___	___ ___

Lesson 7: Name and count shapes as parts of a whole, recognizing relative sizes of the parts.

2. Write the number of equal parts in each shape.

a.	b.	c.
___	___	___
d.	e.	f.
___	___	___

3. Draw one line to make this triangle into 2 equal triangles.

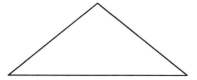

4. Draw one line to make this square into 2 equal parts.

5. Draw two lines to make this square into 4 equal squares.

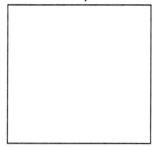

Name _____ Date _____

Circle the shape that has equal parts.

How many equal parts does the shape have? _____

A STORY OF UNITS   Lesson 7 Homework 1•5

Name _____   Date _____

1. Are the shapes divided into equal parts? Write **Y** for yes or **N** for no. If the shape has equal parts, write how many equal parts there are on the line. The first one has been done for you.

a. ⊘ Y 2	b. M ___ ___	c. Y ___ ___
d. ⊘ ___	e. △ ___	f. ◇ ___
g. ▭ ___	h. △ ___	i. ◯ ___
j. △ ___	k. ⏢ ___	l. ⬡ ___
m. ⬡ ___	n. ✦ ___	o. ★ ___

Lesson 7: Name and count shapes as parts of a whole, recognizing relative sizes of the parts.

2. Draw 1 line to make 2 equal parts. What smaller shapes did you make?

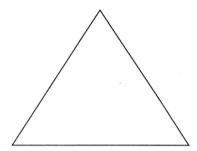

I made 2 _____.

3. Draw 2 lines to make 4 equal parts. What smaller shapes did you make?

I made 4 _____.

4. Draw lines to make 6 equal parts. What smaller shapes did you make?

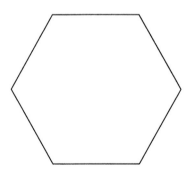

I made 6 _____.

Lesson 8

Objective: Partition shapes and identify halves and quarters of circles and rectangles.

Suggested Lesson Structure

■ Fluency Practice   (15 minutes)
■ Application Problem   (5 minutes)
■ Concept Development   (30 minutes)
■ Student Debrief   (10 minutes)
   Total Time   **(60 minutes)**

Fluency Practice (15 minutes)

- Core Fluency Differentiated Practice Sets **1.OA.6**   (5 minutes)
- 5 More **1.NBT.4**   (5 minutes)
- Make Ten Addition with Partners **1.OA.6**   (5 minutes)

Core Fluency Differentiated Practice Sets (5 minutes)

Materials: (S) Core Fluency Practice Sets (Lesson 3 Core Fluency Practice Sets)

Note: Give the appropriate Practice Set to each student. Help students become aware of their improvement. After students complete today's Practice Sets, ask them to raise their hands if they tried a new level today or improved their scores from the previous day.

Students complete as many problems as they can in 90 seconds. Assign a counting pattern and start number for early finishers, or tell them to practice make ten addition or subtraction on the back of their papers. Collect and correct any Practice Sets completed within the allotted time.

5 More (5 minutes)

Note: This activity prepares students for Lesson 11, where they add 5 minutes until they reach 30 minutes to connect half past the hour to 30 minutes past the hour. The suggested sequence of this activity enables students to use their experience with analogous addition to add 5. Be sure to provide enough think time for students to mentally add or count on, as needed. If students require more support, consider replacing this activity with Whisper Count from Lesson 7.

T: On my signal, say the number that is 5 more. 0. (Pause. Snap.)
S: 5.

T: 10. (Pause. Snap.)
S: 15.

Continue with the following suggested sequence: 20, 30, 5, 15, 25.

Make Ten Addition with Partners (5 minutes)

Materials: (S) Personal white board

Note: This fluency activity reviews how to use the Level 3 strategy of making ten to add two single-digit numbers.

Repeat the activity from Lesson 7.

Application Problem (5 minutes)

Peter and Fran each have an equal number of pattern blocks. There are 12 pattern blocks altogether. How many pattern blocks does Fran have?

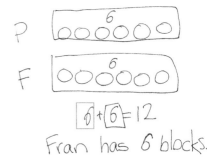

Note: In today's Application Problem, students explore their understanding of the word *equal*. Note the various methods students have for solving the problem. Some of these methods may be useful in supporting students' understanding of equal parts, as applied in today's Concept Development.

Concept Development (30 minutes)

Materials: (T) Example images (Template 1), circles and rectangles (Template 2), projector (S) Circles and rectangles (Template 2), personal white board

Note: The circles and rectangles template should be cut in half. Distribute the top half—images of pizza—to students at the start of the lesson.

Gather students in the meeting area with the circles and rectangles template inserted into their personal white boards.

T: Last night, my brother and I bought a small pizza to share. We agreed we would each eat **half** of the pizza, or one out of two equal parts. My brother cut the pizza for us to share, and it looked like this. (Show Image 1.)

T: Why do you think I was mad? What's wrong with my brother's version of *fair shares*?

S: One piece is much bigger than the other piece. → They are not cut into equal parts.

T: If my brother and I are going to share this pizza fairly, we need to each have an equal part. To have one **half** of the pizza, the two parts need to be the same size. On your personal white boards, draw a line to show how the pizza should have been cut.

S: (Partition the circle into approximately two equal parts.)

T: (Use a student example to share with the class.) Yes! Now, I can get one half of the pizza because each of the two parts is the same size.

Sicilian pizza

T: Sometimes we buy Sicilian pizza, which is shaped like a rectangle. (Project Image 2.) How can we cut this to be in two equal parts, or two **halves of** the pizza? Draw a line on your personal white boards to show how you would cut the rectangular pizza. (Wait as students draw.)

T: I see more than one idea. Who would like to share how he cut the pizza to be two equal parts, two halves of the pizza?

S: I cut the pizza across (horizontally.) → I cut the pizza up and down (vertically.) → I cut the pizza across from one corner to the other (diagonally.)

T: Will my brother and I get the exact same amount to eat?

S: Yes!

T: Wow, we found three different ways to cut the pizza into halves! Good job!

T: I need your help, though, because sometimes our mom and dad eat with us. How can we share that rectangular pizza equally among all four of us?

S: You need to cut it into four pieces. → The pieces need to be the same size. → You can just cut it again the other way. That's what my mom does with my sandwiches!

T: Draw lines to show how you would cut the rectangular pizza so we would have four equal parts.

S: (Students draw lines on their personal white boards over the rectangular pizza.)

T: How did you cut one pizza into four equal parts, or **fourths**?

S: I drew one line up and down (vertically) and the other line across (horizontally). → I drew all my lines in the same way. Everyone would get a strip of pizza that is the same size.

T: Great job! These are all **fourths of**, or **quarters of**, the pizza. It is cut into four pieces that are the same size.

S: I drew two lines diagonally through the middle from each corner. That makes four triangles, but they are not all the same shape, so I wonder if the four pieces are equal shares even though they are not the same shape.

T: Interesting observation. I wonder, too! (While the diagonal cuts *would* create equal shares, the shapes created are not exactly the same. These are the most challenging types of equal parts. Consider exploring cutting shapes diagonally as an extension to the lesson.)

NOTES ON MULTIPLE MEANS OF REPRESENTATION:

Highlight the critical vocabulary for English language learners throughout the lesson by showing object(s) as a visual or gesturing while saying the words. In this lesson, vocabulary to highlight is *half, fourths, quarters, quarter-circle,* and *half-circle.*

Lesson 8: Partition shapes and identify halves and quarters of circles and rectangles.

T: Let's try to make fourths, or quarters, from the circle-shaped pizza. (Observe as students draw lines on their personal white boards. Support students in visually checking that they have four equal parts to their circle.)

T: How did you cut the pizza so that it was cut into four equal parts, which we call fourths, or quarters?

S: I cut across (horizontally) and up and down (vertically). → I tried to cut it in straight lines, like I did with the rectangle, but the end pieces were too small. I had to cut it through the middle to keep the parts the same size.

T: Good observations. Sometimes it's easier to make equal parts by cutting them in particular ways. Can the circle AND the rectangle both be cut into fourths?

S: Yes!

T: So, if there are four people sharing a pizza, whichever shape we're using, we need the whole pizza to be cut into...?

S: Fourths! (Or quarters.)

T: If there are two people sharing, we need the whole pizza to be cut into...?

S: Halves!

T: Look at this shape. (Project Image 3, a quarter-circle.) This shape is called a **quarter-circle**. How do you think it got its name?

Quarter circle

Half circle

S: It comes from a whole circle that got cut into fourths, or quarters. → It comes from a circle cut into four equal parts. → If you put it together with 3 other pieces that are the same size, you would get a whole circle. Four quarters make one whole.

MP.7

T: If this shape (point to Image 3, the quarter-circle) is called a quarter-circle, what do you think this shape is called? (Project Image 4, the half-circle.)

S: A **half-circle**!

T: How did you know?

S: It comes from a whole circle that got cut in half. → It comes from a circle cut into two parts. → If you put it together with another piece that is the same size, you would get a whole circle. Two halves make one whole.

Distribute the bottom half of the circles and rectangles template to be inserted into the personal white boards. Invite students to partition the shapes in halves. Discuss the various positions of their lines and the importance of having equal parts no matter which way the shape is partitioned. Repeat this process having students partition the shapes into fourths, or quarters.

NOTES ON MULTIPLE MEANS OF REPRESENTATION:

Some students may benefit from various aids when modeling halves and fourths. Providing rulers may help students draw straight lines. Other students may need to cut out or fold paper to accurately convey equal partitions.

Problem Set (10 minutes)

Students should do their personal best to complete the Problem Set within the allotted 10 minutes. For some classes, it may be appropriate to modify the assignment by specifying which problems they work on first. Some problems do not specify a method for solving.

A STORY OF UNITS   Lesson 8 1•5

Student Debrief (10 minutes)

Lesson Objective: Partition shapes and identify halves and quarters of circles and rectangles.

The Student Debrief is intended to invite reflection and active processing of the total lesson experience.

Invite students to review their solutions for the Problem Set. They should check work by comparing answers with a partner before going over answers as a class. Look for misconceptions or misunderstandings that can be addressed in the Debrief. Guide students in a conversation to debrief the Problem Set and process the lesson.

Any combination of the questions below may be used to lead the discussion.

- What word did we learn today to help us name the pieces of a shape cut into two equal parts? (**Half** or **halves**.) (Hold up the rectangular pizza image with a line to divide it in half.) How much of the pizza is one part? (**Half of** the pizza.)

- What two different ways can we name the parts of a shape that is cut into four equal parts? (**Fourths** or **quarters**.) (Hold up the rectangular pizza image, divided into quarters.) How much of the pizza is one part? (A **quarter of** the pizza, or a **fourth of** the pizza.) Look at Problem 1. Find an example of a shape that is not divided into halves. How did you decide that the parts were not equal?

- Look at Problem 2. Find an example of a shape that is not divided into quarters. How did you decide it did not have four equal parts?

- (Display the chart created during Lesson 7.) Let's look at the shapes we made with our tangram pieces during our last lesson. Can we name the size of the equal pieces in each of our shapes?

- Someone told me that when you cut rectangles into quarters, you always get smaller rectangles. Is that true? Look over your Problem Set to support your thinking with examples.

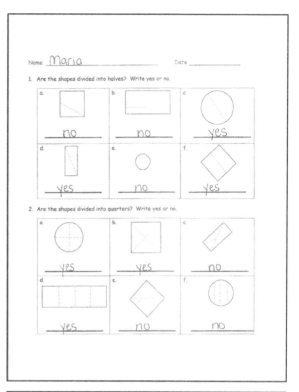

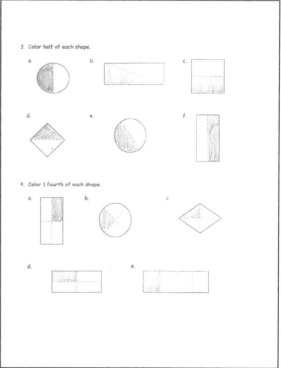

Lesson 8: Partition shapes and identify halves and quarters of circles and rectangles. 115

- What is the shape of a **half-circle**? How does it compare to a **quarter-circle**?
- How many quarter-circles would you need to make a whole circle? How many quarter-circles would you need to make a half-circle? Explain your thinking.
- Think about today's fluency activities. Choose one of the activities we completed, and tell your partner how it can help you practice your number work.

Exit Ticket (3 minutes)

After the Student Debrief, instruct students to complete the Exit Ticket. A review of their work will help with assessing students' understanding of the concepts that were presented in today's lesson and planning more effectively for future lessons. The questions may be read aloud to the students.

Name _____   Date _____

1. Are the shapes divided into halves? Write yes or no.

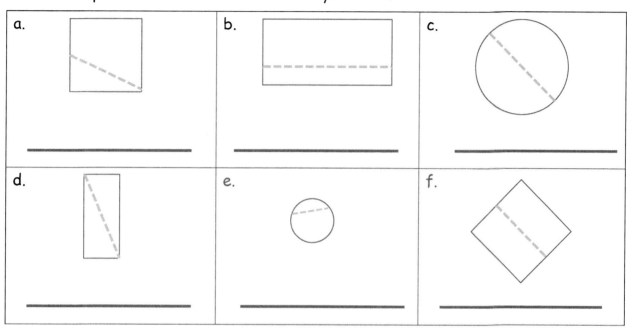

2. Are the shapes divided into quarters? Write yes or no.

3. Color half of each shape.

a. 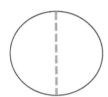   b.    c.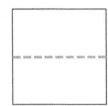

d. 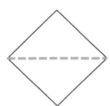   e.    f.

4. Color 1 fourth of each shape.

a. 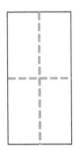   b. 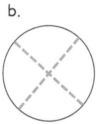   c.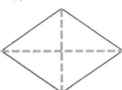

d.    e.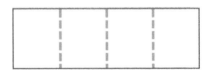

Name _____ Date _____

| Color 1 fourth of this square. | Color half of this rectangle. |
| Color half of this square. | Color a quarter of this circle. |

Lesson 8: Partition shapes and identify halves and quarters of circles and rectangles.

A STORY OF UNITS　　　　　　　　　　　　　　　　　　　　　　　　Lesson 8 Homework 1•5

Name _____   Date _____

1. Circle the correct word(s) to tell how each shape is divided.

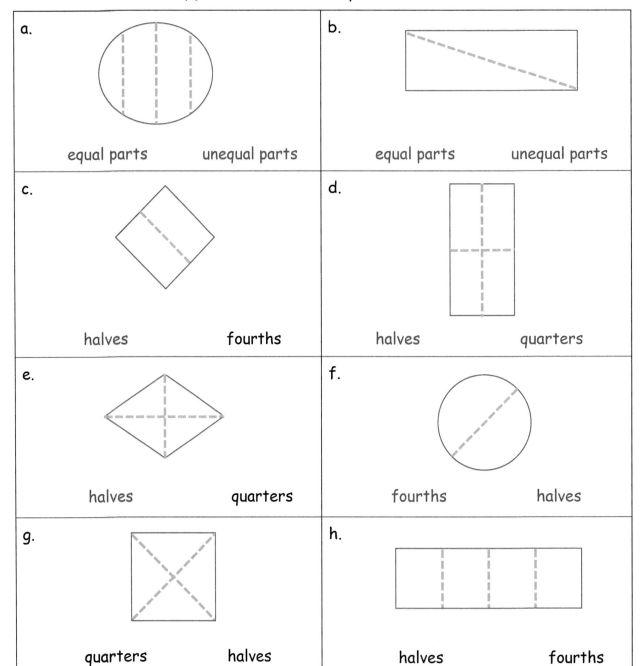

2. What part of the shape is shaded? Circle the correct answer.

a.

   1 half   1 quarter

b.

   1 half   1 quarter

c.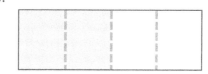

   1 half   1 quarter

d.

   1 half   1 quarter

3. Color 1 quarter of each shape.

4. Color 1 half of each shape.

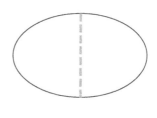 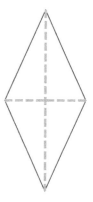

Lesson 8: Partition shapes and identify halves and quarters of circles and rectangles.

Image 1

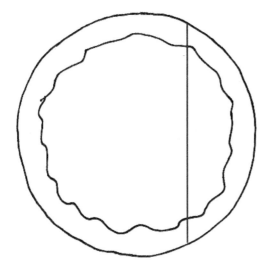

Image 2

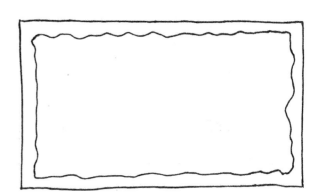

Image 3

Image 4

example images

A STORY OF UNITS — Lesson 8 Template 2 — 1•5

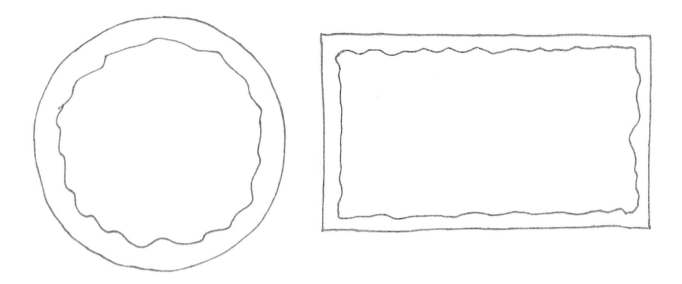

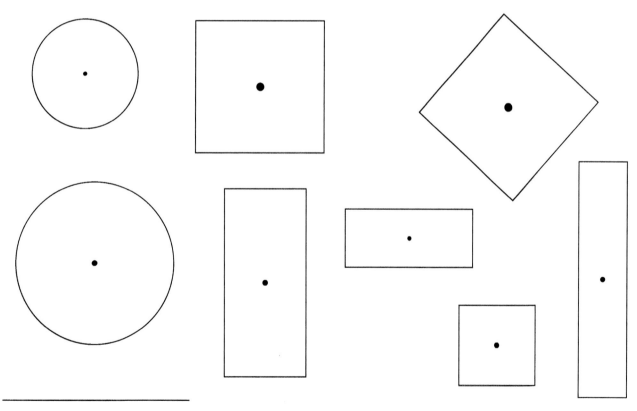

circles and rectangles

Lesson 8: Partition shapes and identify halves and quarters of circles and rectangles.

Lesson 9

Objective: Partition shapes and identify halves and quarters of circles and rectangles.

Suggested Lesson Structure

■ Fluency Practice   (15 minutes)
■ Application Problem (5 minutes)
■ Concept Development (30 minutes)
■ Student Debrief   (10 minutes)
  Total Time   **(60 minutes)**

Fluency Practice (15 minutes)

- Grade 1 Core Fluency Sprint **1.OA.6**   (10 minutes)
- Make It Equal: Addition Expressions **1.OA.6** (5 minutes)

Grade 1 Core Fluency Sprint (10 minutes)

Materials: (S) Core Fluency Sprint (Lesson 1 Core Fluency Sprint)

Note: When choosing a counting sequence to practice between Sides A and B, consider having students whisper count by fives to 30 and back. Although counting by fives is not a Grade 1 standard, in Lesson 11 students will be adding 5 minutes until they reach 30 minutes to build an understanding of half past the hour.

Choose an appropriate Sprint based on the needs of the class. If the majority of students completed the first three quadrants on the previous Sprint, move to the next Sprint listed in the sequence provided below (Core Fluency Sprint List). If many students are not making it to the third quadrant, consider repeating the same Sprint. As students work, pay attention to their strategies and the number of problems they answer to consider for future Sprint administration.

Core Fluency Sprint List:

- Core Addition Sprint 1 (Targets core addition and missing addends.)
- Core Addition Sprint 2 (Targets the most challenging addition within 10.)
- Core Subtraction Sprint (Targets core subtraction.)
- Core Fluency Sprint: Totals of 5, 6, and 7 (Develops understanding of the relationship between addition and subtraction.)
- Core Fluency Sprint: Totals of 8, 9, and 10 (Develops understanding of the relationship between addition and subtraction.)

Make It Equal: Addition Expressions (5 minutes)

Materials: (S) Numeral cards (Lesson 1 Fluency Template), one "=" card, two "+" cards

Note: This activity builds fluency with Grade 1's core addition facts and promotes an understanding of equality. The suggested sets move from simple to complex, so students can progress through them at their own rate.

Assign students partners of equal ability. Students arrange numeral cards from 0 to 10, including the extra 5. Place the "=" card between partners. Write or project the suggested sets. Partners take the numeral cards that match the numbers written to make two equivalent expressions (e.g., 10 + 0 = 5 + 5).

Suggested sets: a) 10, 0, 5, 5   b) 9, 8, 2, 1   c) 3, 6, 4, 7   d) 1, 2, 6, 5
                 e) 1, 2, 5, 4   f) 3, 5, 4, 2   g) 2, 3, 5, 6   h) 3, 4, 5, 6
                 i) 4, 5, 9, 10   j) 9, 3, 2, 8   k) 8, 5, 9, 4   l) 5, 6, 8, 7

Application Problem (5 minutes)

Emi cut a square brownie into fourths. Draw a picture of the brownie. Emi gave away 3 parts of the brownie. How many pieces does she have left?

Extension: What part, or fraction, of the whole brownie is left?

Note: Today's Application Problem provides students with the opportunity to apply the terminology of *fourths*. Students solve the relatively familiar *take away with result unknown* problem type using fractions as a type of unit.

Concept Development (30 minutes)

Materials: (T) Chart paper, 2 pieces of blank paper of the same size (preferably different colors), document camera (S) Pairs of shapes (Template), personal white board

Gather students in the meeting area with the pairs of shapes template inserted into their personal white boards.

- T: Partner A, draw one line to cut your pizza into halves.
- T: Partner B, draw two lines to cut your pizza into quarters.
- T: Who has more slices?
- S: Partner B has more slices. → Partner B has four slices; Partner A only has two slices.
- T: Partner A, color one slice of your pizza. Show me your slice.
- T: Partner B, color one slice of your pizza. Show me your slice.

NOTES ON MULTIPLE MEANS OF ENGAGEMENT:

For kinesthetic learners, it may be beneficial to provide two pieces of blank paper to student partners and have students cut one fourth from one paper and one half from another paper along with the teacher.

Lesson 9: Partition shapes and identify halves and quarters of circles and rectangles.

T: Partners, put your half and your quarter next to each other.
T: Point to the piece of pizza that is larger. Whose piece is larger?
S: Partner A's.
T: Now, look at your whole pizza. Who has a larger number of slices?
S: Partner B has more slices. → Partner B has four slices of pizza. Partner A only has two slices of pizza.
T: Do you want one half of a yummy pizza or one quarter of a yummy pizza? Discuss this with your partner. Explain your choice. (Listen as students share their thinking, and then repeat the question before having students answer.)
S: I want one half of the pizza because a half is larger than one quarter of the pizza. → To get one quarter of the pizza, you have to cut the two halves of the pizza in half again. That's a lot smaller. I would want one half of the pizza. → You need two quarters of the pizza to have the same amount as one half of the pizza.

Draw two circles of equal size on the board. Invite a student volunteer to draw a line to cut the first circle into two halves. Ask the student to color in one half. Label as one half of the circle. Repeat the process with the other circle, coloring in and labeling one fourth of the circle.

T: Let's try that with the rectangles and see if that's still true. This time, I'll use paper to actually cut and compare. Which will be larger, one half of this piece of paper or one fourth of the paper? Talk with your partner, and explain your thinking. (Listen as students share their thinking.)
T: I'm going to fold the paper first to be sure I'm cutting equal parts. (Fold and cut the paper into halves. Ask a student volunteer to hold one half.)
T: How much of the paper is he holding?
S: One half of the paper!
T: Let's cut this same-size piece of paper into four equal parts now, so we can compare one fourth, or one quarter, of the paper with one half of the paper. This time, I'm going to fold the paper in half and then in half again to make four equal parts. (Fold and cut paper into fourths.)
T: Are all of my parts equal?
S: Yes!
T: How much of the paper is each piece?
S: One fourth of the paper! (Or, one quarter of the paper.)
T: (Ask a student volunteer to hold one fourth next to the student who is holding one half.) Which piece is larger, or greater, one half of the paper or one fourth of the paper?
S: One half of the paper!
T: How many pieces did we make when we cut the paper into halves?

NOTES ON
MULTIPLE MEANS
OF REPRESENTATION:

To support students' vocabulary development, write *one fourth of paper* on each of the four equal pieces of paper and *one half of paper* on each of the two equal pieces of paper. These can be posted in the room as a reference, helping students visualize the concept as well as the relationship between concepts.

Lesson 9: Partition shapes and identify halves and quarters of circles and rectangles.

S: Two pieces.
T: How many pieces did we make when we cut the paper into quarters?
S: Four pieces.
T: So, when we cut the paper into two pieces to make halves, our pieces were this size. (Hold up halves.)
T: What happened to the size of our pieces when we cut the same size paper into four pieces to make quarters?
S: The parts became smaller.
T: Why are the pieces smaller now? Talk to your partner.
S: We cut the paper into more pieces. → We have more parts, but each part is smaller. → The piece of paper is the same size, so if you cut it up into more equal parts, the parts will be smaller.
T: On your personal white boards, you have pairs of the same shape. Draw lines and color in one half of the first shape, and then draw lines and color in one quarter, or one fourth, of the other shape. With your partner, see if one fourth is smaller than one half every time or just sometimes.

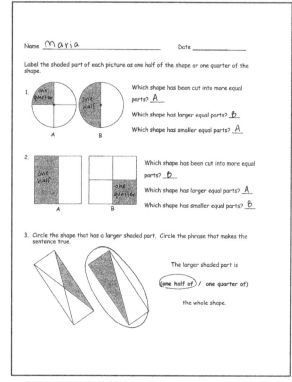

When most students have completed the task, have students show their personal white boards under the document camera and explain their findings.

Problem Set (10 minutes)

Students should do their personal best to complete the Problem Set within the allotted 10 minutes. For some classes, it may be appropriate to modify the assignment by specifying which problems they work on first.

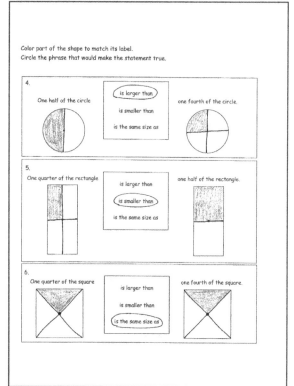

Student Debrief (10 minutes)

Lesson Objective: Partition shapes and identify halves and quarters of circles and rectangles.

The Student Debrief is intended to invite reflection and active processing of the total lesson experience.

Invite students to review their solutions for the Problem Set. They should check work by comparing answers with a partner before going over answers as a class. Look for misconceptions or misunderstandings that can be addressed in the Debrief. Guide students in a conversation to debrief the Problem Set and process the lesson.

Any combination of the questions below may be used to lead the discussion.

- Look at Problem 1. Which shaded part is greater, or larger? Is this true for your other problems? Is one half of a shape always larger than one fourth of the same shape?
- If you want *more* pieces, should you cut your shape into halves or quarters? If you want *larger* pieces, should you cut your shape into halves or quarters? Explain your thinking.
- Why does cutting something into fourths make the equal parts smaller than cutting it into halves?
- Let's think about the first question I asked you today. Would you rather have one half of a yummy pizza or one quarter of a yummy pizza? Explain your thinking. (Choose students who may be better able to express accurate reasoning since participating in the lesson.)
- Look at the Application Problem. Share your drawing with your partner. Did you cut your brownie into quarters in the same way or in a different way? How did you make sure you created four equal parts?

Exit Ticket (3 minutes)

After the Student Debrief, instruct students to complete the Exit Ticket. A review of their work will help with assessing students' understanding of the concepts that were presented in today's lesson and planning more effectively for future lessons. The questions may be read aloud to the students.

Name _____ Date _____

Label the shaded part of each picture as one half of the shape or one quarter of the shape.

1.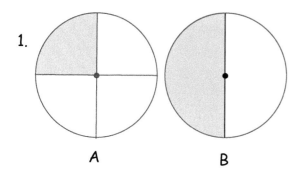

Which shape has been cut into more equal parts? ____

Which shape has larger equal parts? ____

Which shape has smaller equal parts? ____

2.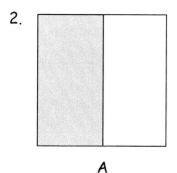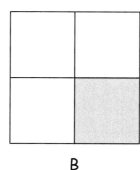

Which shape has been cut into more equal parts? ____

Which shape has larger equal parts? ____

Which shape has smaller equal parts? ____

3. Circle the shape that has a larger shaded part. Circle the phrase that makes the sentence true.

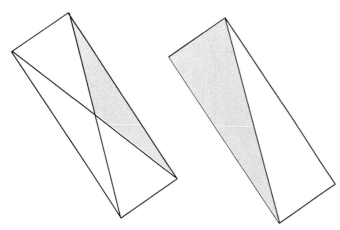

The larger shaded part is

(one half of / one quarter of)

the whole shape.

Color part of the shape to match its label.

Circle the phrase that would make the statement true.

4.

One half of the circle   is larger than   one fourth of the circle.

is smaller than

is the same size as

 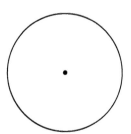

5.

One quarter of the rectangle   is larger than   one half of the rectangle.

is smaller than

is the same size as

6.

One quarter of the square   is larger than   one fourth of the square.

is smaller than

is the same size as

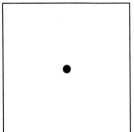 

Name _____ Date _____

1. Circle **T** for true or **F** for false.

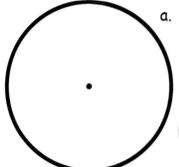

   a. One fourth of the circle is larger than one half of the circle.

   T   F

   b. Cutting the circle into quarters gives you more pieces than cutting the circle into halves.

   T   F

2. Explain your answers using the circles below.

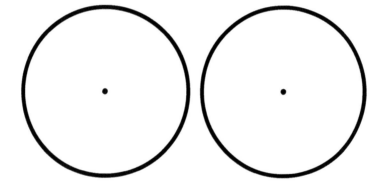

_____

_____

Name _____ Date _____

1. Label the shaded part of each picture as one half of the shape or one quarter of the shape.

   A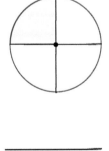

   Which picture has been cut into more equal parts? _____

   Which picture has larger equal parts? _____

   B

   Which picture has smaller equal parts? _____

2. Write whether the shaded part of each shape is a half or a quarter.

   a.

   _____   _____

   b.

   _____   _____

   c.

   _____   _____

   d.

   _____   _____

A STORY OF UNITS   Lesson 9 Homework 1•5

3. Color part of the shape to match its label. Circle the phrase that would make the statement true.

a.

   One quarter of the square   one half of the square.

                              is larger than

                              is smaller than

                              is the same size as

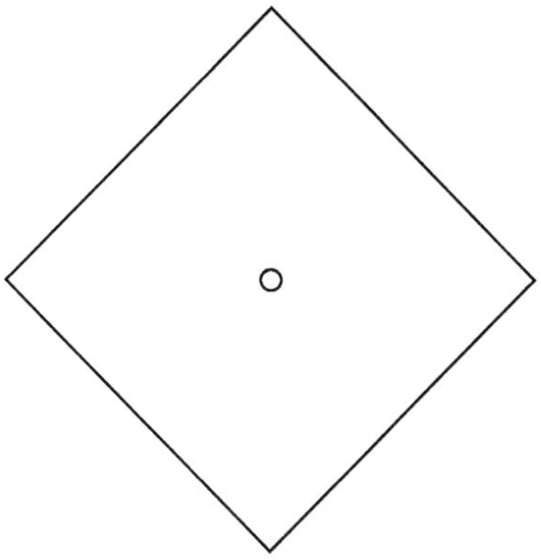

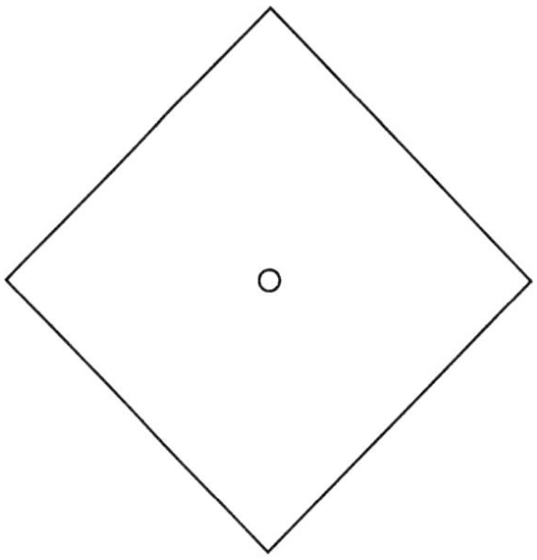

b.

                              is larger than

                              is smaller than
   One quarter of the rectangle   one fourth of the rectangle.
                              is the same size as

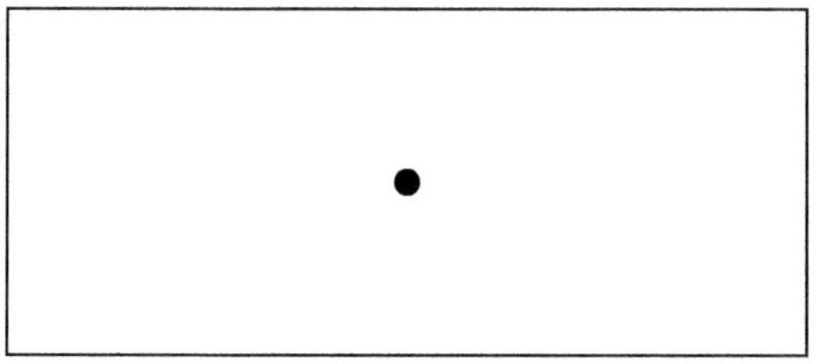

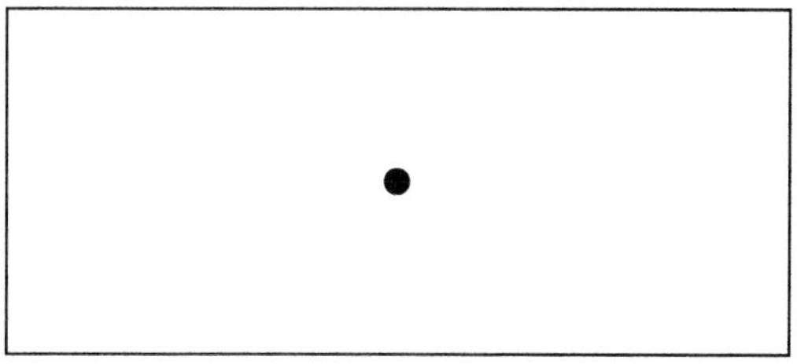

Lesson 9: Partition shapes and identify halves and quarters of circles and rectangles.

133

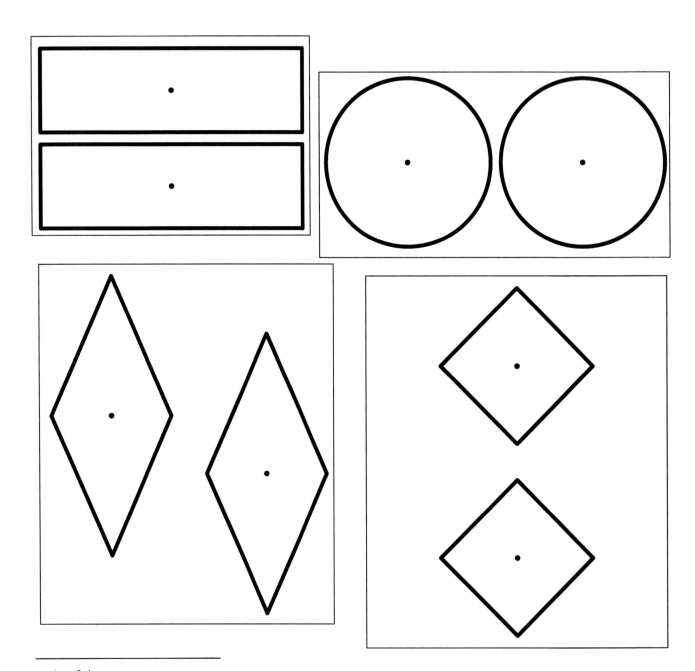

pairs of shapes

A STORY OF UNITS

Mathematics Curriculum

GRADE 1 • MODULE 5

Topic D
Application of Halves to Tell Time

1.MD.3, 1.G.3

Focus Standards:	1.MD.3	Tell and write time in hours and half-hours using analog and digital clocks. Recognize and identify coins, their names, and their values.
	1.G.3	Partition circles and rectangles into two and four equal shares, describe the shares using the words *halves*, *fourths*, and *quarters*, and use the phrases *half of*, *fourth of*, and *quarter of*. Describe the whole as two of, or four of the shares. Understand for these examples that decomposing into more equal shares creates smaller shares.
Instructional Days:	4	
Coherence -Links from:	GK–M2	Two-Dimensional and Three-Dimensional Shapes
-Links to:	G2–M8	Time, Shapes, and Fractions as Equal Parts of Shapes

Topic D builds on students' knowledge of parts of circles to tell time. In Lesson 10, students count and color the parts on a partitioned circle, forming the base of a paper clock. Relating this 12-section circle to the clock, students learn about the hour hand and tell time on both analog and digital clocks.

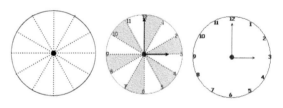

In Lesson 11, students recognize the two half-circles on the circular clock face and connect this understanding with the half hour. Counting by fives to 30, students see that there are two 30-minute parts that make 1 hour, helping them connect the time displayed on a digital clock with the time displayed on an analog clock. Students notice that the hour hand is halfway through, but still within, the hour section on the partitioned paper clock. They tell time to the half hour on both analog and digital clocks.

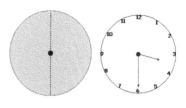

Students continue to practice these skills in Lesson 12. In Lesson 13, they extend these new skills to telling time to the hour and half-hour using a variety of analog and digital clock faces.

A Teaching Sequence Toward Mastery of Application of Halves to Tell Time
Objective 1: Construct a paper clock by partitioning a circle and tell time to the hour. (Lesson 10)
Objective 2: Recognize halves within a circular clock face and tell time to the half hour. (Lessons 11–13)

Lesson 10

Objective: Construct a paper clock by partitioning a circle and tell time to the hour.

Suggested Lesson Structure

- **Fluency Practice** (10 minutes)
- **Application Problem** (5 minutes)
- **Concept Development** (38 minutes)
- **Student Debrief** (7 minutes)
- **Total Time** **(60 minutes)**

Fluency Practice (10 minutes)

- Grade 1 Core Fluency Sprint **1.OA.6** (10 minutes)

Grade 1 Core Fluency Sprint (10 minutes)

Materials: (S) Core Fluency Sprint (Lesson 1 Core Fluency Sprint)

Note: Based on the needs of the class, select a Sprint from Lesson 1. Consider the following options:

1. Re-administer the previous lesson's Sprint.
2. Administer the next Sprint in the sequence.
3. Differentiate. Administer two different Sprints. Simply have one group do a counting activity on the back of the Sprint, while the other group corrects the second Sprint.

Application Problem (5 minutes)

Kim drew 7 circles. Shanika drew 10 circles. How many fewer circles did Kim draw than Shanika?

Note: Students continue to practice the *compare with difference unknown* problem type in today's problem. Children who struggle with this problem type benefit from seeing and hearing their peers' solution strategies. After students describe their solutions, let the class know this is a *compare* problem. Invite students to share why, explaining what is being compared. Module 6 begins with direct instruction on these types of problems. Keep note of which types of problems students are struggling with, as well as which problems they solve successfully. This can assist in targeting

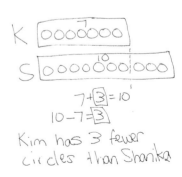

instruction at the start of the next module. Circles were chosen as the context for the problem because of their connection to today's Concept Development.

Concept Development (38 minutes)

Materials: (T) Partitioned circle (Template 1), digital clock (Template 2) (S) Partitioned circle (Template 1) printed on cardstock, scissors, pencil, yellow crayon, orange crayon, brad fastener, personal white board

NOTES ON MULTIPLE MEANS OF ENGAGEMENT:

Students who struggle with fine-motor cutting skills would benefit from using a pre-cut circle. Have some ready for these students to use during the lesson.

Note: Before the lesson, cut off the bottom section of the partitioned circle templates so that clock hands can be distributed later in the lesson. (Precut the teacher's set of clock pieces for ease of use during the lesson.)

Distribute the top section of the partitioned circle template, along with pencils and scissors, to students seated at desks or tables.

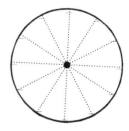

- T: What shape is on this paper?
- S: A circle!
- T: Cut out the circle. Use careful eyes and careful fingers because we will be using this circle for the next three days. Only cut the dark, bold line that forms the circle. (Hold up the circle as a demonstration.)
- S: (Cut out the circle.)
- T: What do you notice about the dotted lines on the circle?

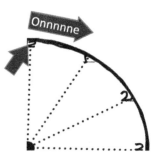

- S: The lines start in the middle and go out to the edge. → There are 12 of them. → No, there are 6, and they all go through the dot in the middle. → They all look equal. The spaces between the lines are about the same size.
- T: (Put the circle under the document camera.) Let's look at the spaces between the lines. Are the parts equal, or are all of the parts different sizes?
- S: The parts are all equal.
- T: Let's count the parts. Let's use our finger to trace the edge as we count. We'll stretch out the counting numbers as we trace the part. When we get to the next piece, we stop and get ready to say the next number. Let me show you.

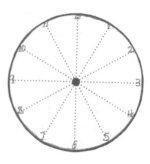

- T: (Trace the edge of the circle under the document camera as students do the same at their seats, while in unison counting the pieces: Ooooonnnnnne! Twoooooo! Etc.)
- S/T: (Repeat together on students' clocks.)

Lesson 10: Construct a paper clock by partitioning a circle and tell time to the hour.

T: How many equal parts do we have?

S: 12 parts!

T: We're going to color in each of the parts, but first, let's use our pencil to trace the edge. We'll trace the edge with brown, and just as we get to the end of the part, or section, we'll put in the number. Watch me. (Start at the edge of the circle, at one dotted line, and trace the edge with the brown colored pencil until reaching the next line. Then, write 1 just before the line, as shown in the image down and to the right. While drawing the line, stretch out saying the word *one*, "Oooooonnnnnne!") Now, you draw a brown line on the edge of your first section, or part, and when you finish saying, "Ooonnne!," write 1 just before the next line. (Point to the sample under the document camera.)

NOTES ON MULTIPLE MEANS OF ENGAGEMENT:

Students may have some difficulty writing the numbers correctly on their clocks. For students who are likely to turn the circle as they write, tape the circle to their desks.

S: (Trace the edge, and number each line as shown. Then, touch and count the parts once more after the numbers are labeled.)

T: Does this look like something you have seen before? Perhaps something we have in our classroom?

S: A clock!

T: Yes, we are making a clock!

T: How many equal parts are labeled on a clock?

S: 12 parts!

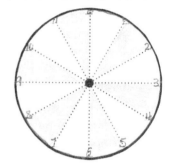

T: Let's color in the 12 parts so we can see them more easily. Alternate between yellow and orange, so each part stands out. Watch as I start the first one. (Color the section between 12 and 1 in yellow, as represented to the right by the lighter shaded section. Then, color the section between 1 and 2 in orange, as represented by the darker shaded section.)

S: (Color the sections.)

T: Look at the clock in our classroom. What else does it have that we need to add to our clocks?

S: Those black things. → There's a red one, too.

T: Those are called clock hands. The red hand is called a second hand, but we are only going to add the black hands for now. The short one is called the **hour hand**, and the longer hand is called the **minute hand**.

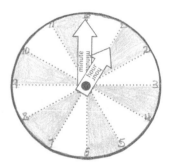

T: (Show the paper hands to the class.) You will cut out your hour hand and minute hand, and push a brad fastener through the dots in each of the three pieces so that the hands are attached to the clock. (Demonstrate and then distribute the paper clock hands.)

S: (Complete the process of making the paper clocks. Position both clock hands pointing toward the 12.)

Lesson 10: Construct a paper clock by partitioning a circle and tell time to the hour.

T: (Show the clock.) This is 12 **o'clock**. At midnight, or 12 o'clock, every night, we begin a new day.

T: As each minute goes by, both hands of the clock move. When the minute hand gets back to the top, and the hour hand reaches the next number, it means we just completed a full hour. (Position the clock hands so that they are set at 1:00.) We can look at the hour hand to tell us which hour we have completed in the new day. This clock's hour hand is now at…?

S: 1.

T: When we get through a full hour, but no extra minutes have passed, we say "o'clock" at the end. What time does this clock read?

S: 1 o'clock!

T: (Show *1:00* using the digital clock template under the document camera.) This is how we see 1 o'clock on a **digital clock**, the kind of electronic clock you see on a microwave, an oven, a cell phone, or a computer. We see the hour first (point to the 1). No extra minutes have passed (point to the zeros).

T: (Position the clock hands so that they are set to 3:00.) What time is this?

S: 3 o'clock!

T: (Show 3:00 using the digital clock template under the document camera.) Three (point to the 3 on the digital clock) o'clock (point to the two zeros).

T: Move the hands of your clock so that it says 11 o'clock. (Wait as students adjust clock hands.)

T: Which hand did you move? The hour hand or the minute hand?

S: The hour hand.

T: To what number is the hour hand pointing?

S: 11.

T: To what number is the minute hand still pointing?

S: 12.

T: Great job! What do you think the digital clock looks like when it reads 11 o'clock?

S: 11, 0, 0.

T: (Show 11:00, using the digital clock template under the document camera.) That's correct!

T: With your partner, choose a time to make on your paper clock by moving just your hour hand. Then, on your personal white board, write the same time the way you would see it on a digital clock.

As students work with a partner, circulate and support student understanding as needed.

Note: The clocks can be collected and redistributed each day during Topic D lessons. Another clock face with numbers already included is provided in Lesson 11 for any students who need a new clock for the upcoming lessons. Alternatively, commercially produced student clocks may be used for Lessons 11–13.

Problem Set (10 minutes)

Students should do their personal best to complete the Problem Set within the allotted 10 minutes. For some classes, it may be appropriate to modify the assignment by specifying which problems they work on first.

Student Debrief (7 minutes)

Lesson Objective: Construct a paper clock by partitioning a circle and tell time to the hour.

The Student Debrief is intended to invite reflection and active processing of the total lesson experience.

Invite students to review their solutions for the Problem Set. They should check work by comparing answers with a partner before going over answers as a class. Look for misconceptions or misunderstandings that can be addressed in the Debrief. Guide students in a conversation to debrief the Problem Set and process the lesson.

Any combination of the questions below may be used to lead the discussion.

- Look at Problem 2. Where did you put the hour hand to show 3 **o'clock**? Is the placement of the hour hand just before, just after, or straight toward the 3? How does your hour hand look different from the minute hand?
- Look at Problem 3. Which times were the easiest for you to read? Why? Which time was the trickiest for you to read? What was tricky about it?
- What is the same about all of the times on your Problem Set? When a new hour has started, and no new minutes have passed since the hour started, which number will the minute hand be pointing toward?
- Besides our classroom, where else have you seen a clock, including a **digital clock**?
- Name the parts of the clock we learned about today. (**Hour hand**, **minute hand**.)
- What is your favorite fluency activity and why? How does that activity help you?

Exit Ticket (3 minutes)

After the Student Debrief, instruct students to complete the Exit Ticket. A review of their work will help with assessing students' understanding of the concepts that were presented in today's lesson and planning more effectively for future lessons. The questions may be read aloud to the students.

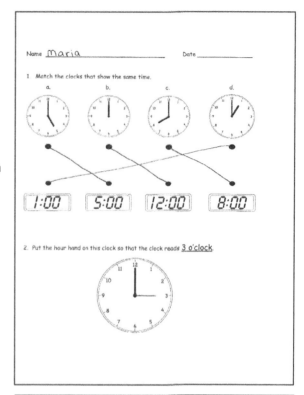
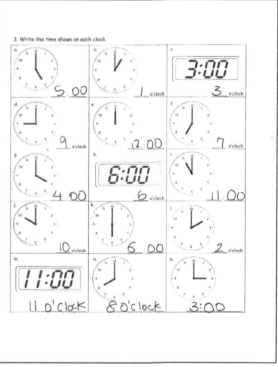

Name _____   Date _____

1. Match the clocks that show the same time.

   a.   b.   c.   d.

   ●   ●   ●   ●

   ●   ●   ●   ●

  1:00   5:00   12:00   8:00

2. Put the hour hand on this clock so that the clock reads 3 o'clock.

Lesson 10 Problem Set

3. Write the time shown on each clock.

Name _____ Date _____

Write the time shown on each clock.

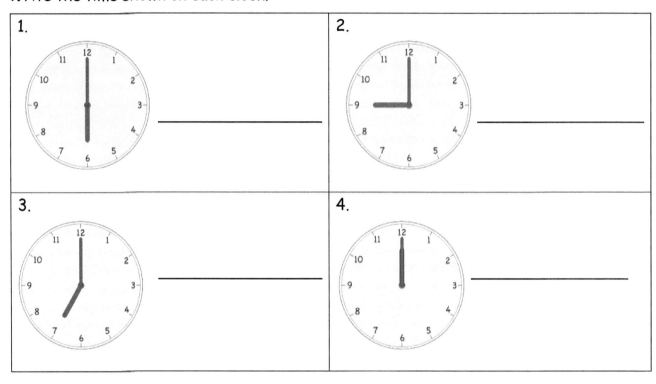

1. _____

2. _____

3. _____

4. _____

Name _____ Date _____

1. Match each clock to the time it shows.

a.

b.

| 4 o'clock |

c.

| 7 o'clock |

d.

| 11 o'clock |

e.

| 10 o'clock |

| 3 o'clock |

f.

| 2 o'clock |

Lesson 10: Construct a paper clock by partitioning a circle and tell time to the hour.

2. Put the hour hand on the clock so that the clock matches the time. Then, write the time on the line.

a.    6 o'clock

b.    9 o'clock   _____

c.    12 o'clock   _____

d.    7 o'clock   _____

e.    1 o'clock   _____

A STORY OF UNITS Lesson 10 Template 1 1•5

_____
partitioned circle

Lesson 10: Construct a paper clock by partitioning a circle and tell time to the hour.

digital clock

Lesson 11

Objective: Recognize halves within a circular clock face and tell time to the half hour.

Suggested Lesson Structure

- ■ Fluency Practice   (14 minutes)
- ▨ Application Problem   (5 minutes)
- ☐ Concept Development   (31 minutes)
- ■ Student Debrief   (10 minutes)
- **Total Time**   **(60 minutes)**

Fluency Practice (14 minutes)

- Core Fluency Differentiated Practice Sets **1.OA.6**   (5 minutes)
- Happy Counting **1.NBT.1**   (2 minutes)
- Think Count **1.OA.5**   (2 minutes)
- Take from Ten Subtraction with Partners **1.OA.6**   (5 minutes)

Core Fluency Differentiated Practice Sets (5 minutes)

Materials: (S) Core Fluency Practice Sets (Lesson 3 Core Fluency Practice Sets)

Note: Give the appropriate Practice Set to each student. Students who completed all of the questions correctly on their most recent Practice Set should be given the next level of difficulty. All other students should try to improve their scores on their current levels.

Students complete as many problems as they can in 90 seconds. Assign a counting pattern and start number for early finishers, or tell them to practice make ten addition or subtraction on the back of their papers. Collect and correct any Practice Set completed within the allotted time.

Happy Counting (2 minutes)

Note: In the next module, students learn addition and subtraction within 100 and extend their counting and number writing skills to 120. Give students practice counting by ones and tens within 100 to prepare them for Module 6. When Happy Counting by ones, spend more time changing directions where changes in tens occur, which is typically more challenging. Happy Count by ones the regular way and the Say Ten way between 40 and 100. Then, Happy Count by tens.

T: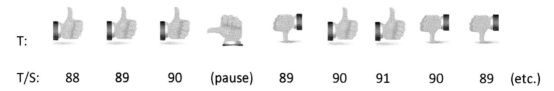

T/S: 88   89   90   (pause)   89   90   91   90   89   (etc.)

Think Count (2 minutes)

Materials: (T) Chart of numbers to 30 with multiples of 5 circled

Note: This activity prepares students for today's lesson, when they will be adding 5 minutes until they reach 30 minutes to connect half past the hour to 30 minutes past the hour.

Display the chart. Students think-count to 20, saying multiples of 5 aloud. Hide the chart, and let students try to remember the sequence, counting slowly by fives to 20. Repeat think-counting and slowly skip-counting first to 25 and then to 30.

Take from Ten Subtraction with Partners (5 minutes)

Materials: (S) Personal white board

Note: This fluency activity reviews how to use the Level 3 strategy of taking from ten when subtracting from teen numbers.

- Assign partners of equal ability.
- Partners choose a minuend for each other between 10 and 20.
- On their personal white boards, students subtract 9, 8, and 7 from their number. Remind students to write the two number sentences (e.g., to solve 13 – 8, they write 10 – 8 = 2, 2 + 3 = 5).
- Partners then exchange personal white boards and check each other's work.

```
13 - 9 = 4      13 - 8 = 5      13 - 7 = 6
  ∧                ∧                ∧
10  3           10  3           10  3

10 - 9 = 1      10 - 8 = 2      10 - 7 = 3
 1 + 3 = 4       2 + 3 = 5       3 + 3 = 6
```

Application Problem (5 minutes)

Tamra has 7 digital clocks in her house and only 2 circular or analog clocks. How many fewer circular clocks does Tamra have than digital clocks? How many clocks does Tamra have altogether?

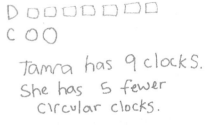

Tamra has 9 clocks.
She has 5 fewer circular clocks.

Note: Today's problem presents both a *put together with result unknown* problem type and a *compare with difference unknown* problem type. Presenting both problems within the same context can support recognizing the differences between the two problem types.

A STORY OF UNITS   Lesson 11 1•5

Concept Development (31 minutes)

Materials: (T) Paper clock created during Lesson 10, document camera, personal white board, dry erase marker, large instructional clock with gears (if available) (S) Paper clock created in Lesson 10 or commercial student clocks, personal white board

NOTES ON MULTIPLE MEANS OF REPRESENTATION:

Consider students' visual needs when determining the appropriate clock for each student. Some students may benefit from large numbers. Other students may find large numbers challenging for identifying the space halfway between the numbers when positioning hands to show the half hour.

Note: For students who need a new paper clock, an additional paper clock with numbers is provided at the end of this lesson (Template).

Distribute materials to students seated at their tables or desks.

- T: In the previous lesson, we read the time when we had whole hours with no extra minutes past the hour. Let's start at 12 o'clock. Where is the minute hand?
- S: At the 12.
- T: Where is the hour hand?
- S: At the 12.
- T: (Position the minute hand on the paper clock accordingly.) When the minute hand moves all the way around the clock, it has been 60 **minutes**, or 1 **hour**. When 1 hour passes, we will be at…?
- S: 1 o'clock!
- T: Which clock hand do we move to show 1 o'clock?
- S: The hour hand. It's the short one.
- T: (Have students count chorally with the teacher, who moves from 1 o'clock, to 2 o'clock, and then 3 o'clock. Move the minute hand all the way around the clock for each hour to show that by moving the minute hand, the hour hand moves to the next hour when the minute hand makes it around the clock once.)
- T: (Draw 3 o'clock, as shown to the right.) How would this look on a digital clock? (Have a student volunteer add the digital time, 3:00, as shown.)
- T: If we were halfway through the next hour, the hour hand would need to be halfway between 3 and…?
- S: 4.
- T: (Position the hour hand halfway between 3 and 4.)
- T: Now, let's think about the minute hand. It would go halfway around the circle. Think about our half circles. Where would we need to stop the minute hand so that it would have traveled across the shape of a half of the circle? Talk with a partner. (Provide students time to discuss.)
- T: (Insert the clock into the personal white board. Starting at the 12, begin to color over each partition of the clock.) Tell me when I have colored half of the clock. Think about the shape of a half circle.

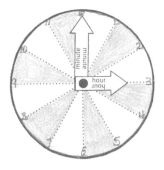

3 o'clock
3:00

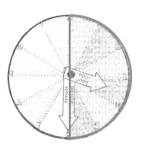

Lesson 11: Recognize halves within a circular clock face and tell time to the half hour.

S: (When the 6 is reached...) STOP!

T: Which number is halfway around the clock?

S: The 6.

T: (Move the minute hand so that it points to the 6.) Yes, if the minute hand were halfway between one hour and another hour, it would be pointing to the 6. We call this time **half past** 3 because it is half an hour past 3 o'clock.

T: Let's see how many minutes are in this half of the hour. We can count each minute, using the little marks on the side of the clock, but it'll be faster to count by groups of 5 minutes, like we do when we whisper count. There are 5 minutes from one number to the next number. (Point to the number 12 on the clock, and then sweep a finger to the number 1 on the clock.)

T: Think about the whisper counting we practiced during Fluency Practice. Count with me, and use your pencil to write the number of minutes next to each dot as we go. (Move a finger along the edge of the clock while counting.) 5...10...15...20...25...30. When the minute hand gets to halfway around and lands on the 6, it has been...?

S: 30 minutes!

T: Another way to say half past 3 is 3:30 because it's 3 hours and 30 minutes since 12 o'clock, when we either started a new day or when we started the afternoon. On a digital clock, half past 3 would look like this. (Write 3:30 on paper. Write *half past 3* next to it.)

T: What time is this? (Point to 3:30.)

S: 3:30.

T: What's another way we can say that it's 3:30?

S: Half past 3.

T: Look at our two clocks. One clock shows 3 o'clock. The other clock shows half past 3, or 3:30. Compare them. What do you notice?

S: The clock on 3 o'clock has its minute hand on 12, and the clock at 3:30 has its minute hand at 6. → The hour hand is pointing directly to 3 on the clock that shows 3 o'clock. The hour hand is pointing between 3 and 4 on the clock that shows 3:30.

> **NOTES ON MULTIPLE MEANS OF ENGAGEMENT:**
>
> If most of the class has difficulty counting by fives, choose to use the hash marks on a commercial teacher clock or student clock, and count by ones to 30.

Half past 3 o'clock
3:30
Three-thirty

Repeat the process of naming a time and having students create the time on their student clocks and then writing the digital time on their personal white boards. Use the following suggested sequence:

- Half past 4
- 10:30
- Half past 11
- Half past 12
- 6:30

A STORY OF UNITS

Lesson 11 1•5

Problem Set (10 minutes)

Students should do their personal best to complete the Problem Set within the allotted 10 minutes. For some classes, it may be appropriate to modify the assignment by specifying which problems they work on first.

Student Debrief (10 minutes)

Lesson Objective: Recognize halves within a circular clock face and tell time to the half hour.

The Student Debrief is intended to invite reflection and active processing of the total lesson experience.

Invite students to review their solutions for the Problem Set. They should check work by comparing answers with a partner before going over answers as a class. Look for misconceptions or misunderstandings that can be addressed in the Debrief. Guide students in a conversation to debrief the Problem Set and process the lesson.

Any combination of the questions below may be used to lead the discussion.

- Look at Problem 4. Which clock shows half past 12 o'clock? Explain your thinking. Remember to use *hour hand* and *minute hand* in your explanation.
- How many minutes are in half an hour? When it is **half past** seven, how many minutes have there been since 7 o'clock? (Extension: If there are 30 minutes in half an hour, how many minutes are in a whole hour?)
- (Write 7:30 on the board.) What are the two ways to say this time?
- When we go around a circle in this direction (motion in a clockwise path), we say we are going *clockwise*. How can knowing about how clocks work help us understand the direction of *clockwise*?
- Look at the Application Problem. What kinds of clocks do you have in your home? Compare the clocks in your home with Tamra's clocks. Who has more clocks? How many more clocks does that person have?

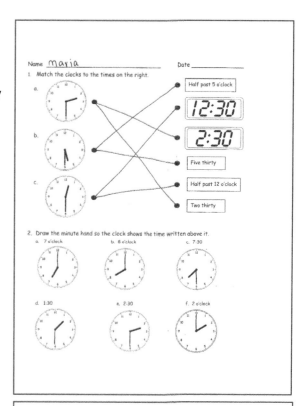

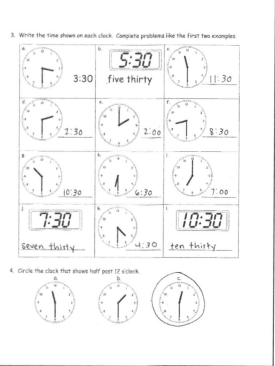

Lesson 11: Recognize halves within a circular clock face and tell time to the half hour.

153

Exit Ticket (3 minutes)

After the Student Debrief, instruct students to complete the Exit Ticket. A review of their work will help with assessing students' understanding of the concepts that were presented in today's lesson and planning more effectively for future lessons. The questions may be read aloud to the students.

Name _____ Date _____

1. Match the clocks to the times on the right.

a.  ●

● Half past 5 o'clock

● 12:30

● 2:30

b.  ●

● Five thirty

c.  ●

● Half past 12 o'clock

● Two thirty

2. Draw the minute hand so the clock shows the time written above it.

a. 7 o'clock   b. 8 o'clock   c. 7:30

d. 1:30   e. 2:30   f. 2 o'clock

Lesson 11: Recognize halves within a circular clock face and tell time to the half hour.

3. Write the time shown on each clock. Complete problems like the first two examples.

4. Circle the clock that shows half past 12 o'clock.

a.    b.    c.

Name _____ Date _____

Draw the minute hand so the clock shows the time written above it.

1. 9:30

2. 3:30

3. Write the correct time on the line.

 _____

A STORY OF UNITS

Lesson 11 Homework 1•5

Name _____ Date _____

Circle the correct clock.

1. Half past 2 o'clock

a.  b.  c.

2. Half past 10 o'clock

a.  b.  c.

3. 6 o'clock

a.  b.  c.

4. Half past 8 o'clock

a.  b.  c.

Lesson 11: Recognize halves within a circular clock face and tell time to the half hour.

Write the time shown on each clock to tell about Lee's day.

5.
Lee wakes up at _____.

6.
He takes the bus to school at _____.

7.
He has math at _____.

8.
He eats lunch at _____.

9.
He has basketball practice at _____.

10.
He does his homework at _____.

11.
He eats dinner at _____.

12.
He goes to bed at _____.

Lesson 11: Recognize halves within a circular clock face and tell time to the half hour.

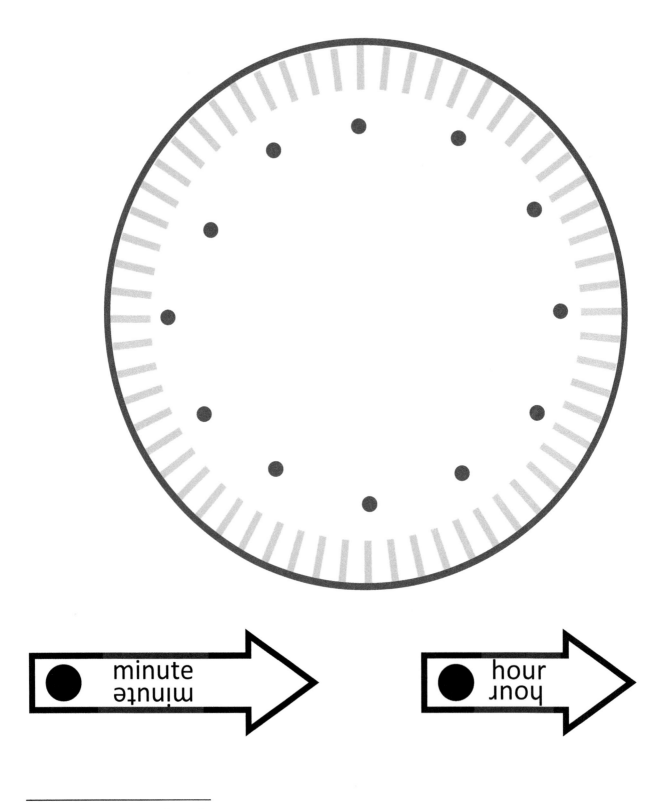

additional paper clock with numbers

Lesson 12

Objective: Recognize halves within a circular clock face and tell time to the half hour.

Suggested Lesson Structure

- ■ Fluency Practice   (15 minutes)
- ▫ Application Problem   (5 minutes)
- ▫ Concept Development   (30 minutes)
- ■ Student Debrief   (10 minutes)
  Total Time   **(60 minutes)**

Fluency Practice (15 minutes)

- Core Fluency Differentiated Practice Sets **1.OA.6**   (5 minutes)
- Happy Counting **1.NBT.1**   (2 minutes)
- Analogous Addition and Subtraction **1.OA.6, 1.NBT.2**   (3 minutes)
- Take from Ten Subtraction with Partners **1.OA.6**   (5 minutes)

Core Fluency Differentiated Practice Sets (5 minutes)

Materials: (S) Core Fluency Practice Sets (Lesson 3 Core Fluency Practice Sets)

Note: Give the appropriate Practice Set to each student. Help students become aware of their improvement. After students do today's Practice Sets, ask them to raise their hands if they tried a new level today or improved their score from the previous day.

Students complete as many problems as they can in 90 seconds. Assign a counting pattern and start number for early finishers, or tell them to practice make ten addition or subtraction on the back of their papers. Collect and correct any Practice Set completed within the allotted time.

Happy Counting (2 minutes)

Note: This activity prepares students for Module 6 by providing practice counting by ones and tens within 100.

Repeat the activity from Lesson 11.

Analogous Addition and Subtraction (3 minutes)

Note: This activity practices Grade 1's core fluency and reminds students to use their knowledge of sums and differences within 10 (e.g., 5 + 3 = 8) to solve analogous problems within 40 (e.g., 15 + 3 = 18, 25 + 3 = 28, and 35 + 3 = 38).

T: On my signal, say the equation with the answer. 6 + 2 = ___. (Pause. Snap.)
S: 6 + 2 = 8.
T: 16 + 2 = ___. (Pause. Snap.)
S: 16 + 2 = 18.

Continue with 26 + 2 and 36 + 2. Then repeat, beginning with other addition or subtraction sentences within 10.

Suggested sequence:

- 5 + 3, 15 + 3, 25 + 3, 35 + 3
- 5 + 4, 4 + 5, 14 + 5, 24 + 5
- 7 + 2, 2 + 7, 12 + 7, 32 + 7

- 6 – 3, 16 – 3, 26 – 3, 36 – 3
- 8 – 2, 18 – 2, 28 – 2, 38 – 2
- 9 – 3, 19 – 3, 29 – 3, 39 – 3

Take from Ten Subtraction with Partners (5 minutes)

Materials: (S) Personal white board

Note: This fluency activity reviews how to use the Level 3 strategy of taking from ten when subtracting from teen numbers.

Repeat activity from Lesson 11.

Application Problem (5 minutes)

Shade the clock from the start of a new hour through half an hour. Explain why that is the same as 30 minutes.

Note: Before beginning today's Concept Development, students have the opportunity to demonstrate their understanding using words and pictures. Circulate, and notice the areas where students are using clear, precise language, as well as elements of their explanation that can use stronger or clearer language. Throughout today's Concept Development, take care to emphasize or extend the lesson around these areas.

Concept Development (30 minutes)

Materials: (T) Instructional clock, paper with quarter of the page cut out to cover the minute hand (see Sequence C figure) (S) Student clock

This lesson is designed to support student understanding of telling time to the half hour. Below are four sequences of problems that can be used, from simple to complex:

- Sequence A reinforces time to the hour.
- Sequence B reinforces discriminating between time to the hour and the half hour.
- Sequence C focuses on positioning the hour hand when telling time to the half hour.
- Sequence D challenges students beyond the standard to apply their ability of telling time to the hour and half hour to story problems.

Choose the sequence that is most appropriate for students. If appropriate, only use part of a sequence.

Sequence A

T: Write the time that matches this clock. (Hold up a clock showing the following times.)
- 11:00
- 2:00
- 6:00

T: On your clock, show the following time. Then, write the time the way it would appear on a digital clock. (Say the following times.)
- 7:00
- 8 o'clock
- 12 o'clock
- 5:00

Sequence B

T: Write the time that matches this clock. (Hold up a clock showing the following times.)
- 7:00
- 12:30 (Ask for both ways to say this time.)
- 1:30 (Ask for both ways to say this time.)

T: On your clock, show the following time. Then, write the time the way it would appear on a digital clock. (Say the following times.)
- Half past 8
- 9:00
- half past 9

NOTES ON MULTIPLE MEANS OF ENGAGEMENT:

While teaching, be sure to provide cross-curricular connections for students. Visit the school or local library to check out books on time to supplement learning during the last two lessons of the module.

- 11:00
- 10:30
- 7 o'clock
- 6:30

Sequence C

T: I'm going to cover the minute hand on this clock. Look closely at the hour hand to decide what time it is. Show the correct time on your clock, and write the time on your personal white board.
(For each time below, cover as much of the clock as possible while showing the hour hand. Place the hour hand directly on the given hour, or halfway between the two numbers, depending on the appropriate position for the given time.)

- 2:00
- 2:30 (Ask for both ways to say this time.)
- 4:00
- 4:30 (Ask for both ways to say this time.)
- 9:30 (Ask for both ways to say this time.)
- 7:30 (Ask for both ways to say this time.)
- 3:00

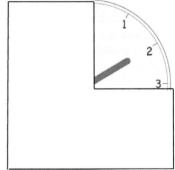

Sequence D

T: Listen to my story, and see if you can determine the time.

- Kim's dance class starts at 3 o'clock. The class lasts half an hour. What does the clock look like when the class ends? Show the time by using your paper clock and writing on your personal white board.
- When I left the house to buy groceries, the clock looked like this. (Show clock at 5:00.) It took me 1 hour to buy groceries and come home. What time did the clock show when I arrived home? Use your paper clock and your personal white board to show the time.
- School begins at 8:30. We have lunch after 3 hours. What time do we have lunch? Use your paper clock and your personal white board to show the time.

NOTES ON MULTIPLE MEANS OF ENGAGEMENT:

Remember to provide challenging extensions for students working above grade level. Giving problems such as those in Sequence D allows students to think about elapsed time. After completing this sequence, advanced students can write their own elapsed time problems to provide another extension to their learning.

For each problem situation, invite students to share how they solved the problem and share the position of the hands on their clock and the time displayed on a digital clock.

Problem Set (10 minutes)

Students should do their personal best to complete the Problem Set within the allotted 10 minutes. For some classes, it may be appropriate to modify the assignment by specifying which problems they work on first.

A STORY OF UNITS

Lesson 12 1•5

Student Debrief (10 minutes)

Lesson Objective: Recognize halves within a circular clock face and tell time to the half hour.

The Student Debrief is intended to invite reflection and active processing of the total lesson experience.

Invite students to review their solutions for the Problem Set. They should check work by comparing answers with a partner before going over answers as a class. Look for misconceptions or misunderstandings that can be addressed in the Debrief. Guide students in a conversation to debrief the Problem Set and process the lesson.

Any combination of the questions below may be used to lead the discussion.

- Look at Problem 1. How did you choose the correct clock? Demonstrate how you know A is the correct answer.
- Look at Problem 4. What is another way to say 9:30? Why is 9:30 also known as *half past 9*?
- Look at Problem 7. How did you draw the clock hands for 12:30? Explain why you placed the minute hand and the hour hand in each location.
- Look at the clock in our room. Is the time closest to a new hour or closest to half past the hour? What time is it right now?
- How could your fluency activities today help you with your subtraction?

Exit Ticket (3 minutes)

After the Student Debrief, instruct students to complete the Exit Ticket. A review of their work will help with assessing students' understanding of the concepts that were presented in today's lesson and planning more effectively for future lessons. The questions may be read aloud to the students.

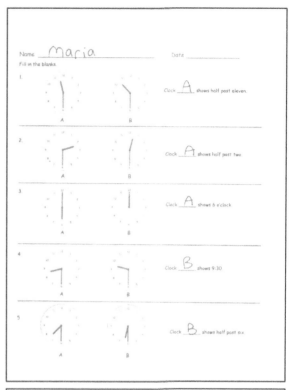

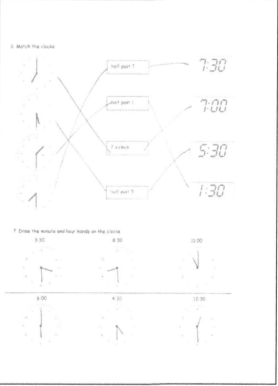

Lesson 12: Recognize halves within a circular clock face and tell time to the half hour.

165

Name _____   Date _____

Fill in the blanks.

1.     Clock _____ shows half past eleven.

2.     Clock _____ shows half past two.

3.     Clock _____ shows 6 o'clock.

4.     Clock _____ shows 9:30.

5.     Clock _____ shows half past six.

6. Match the clocks.

a.    half past 7

b.    half past 1

c.    7 o'clock

d.    half past 5

7. Draw the minute and hour hands on the clocks.

a. 3:30

b. 8:30

c. 11:00

d. 6:00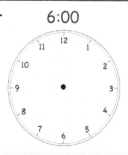

e. 4:30

f. 12:30

Lesson 12: Recognize halves within a circular clock face and tell time to the half hour.

Name _____ Date _____

Draw the minute and hour hands on the clocks.

1. 1:30

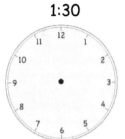

2. 10:00

3. 5:30

4. 7:30

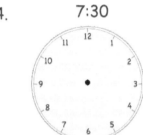

Name _____ Date _____

Write the time shown on the clock, or draw the missing hand(s) on the clock.

1.	10 o'clock	2.	half past 10 o'clock
3.	8 o'clock	4.	_____
5.	3 o'clock	6.	half past 3 o'clock
7.	_____	8.	half past 6 o'clock
9.	half past 9 o'clock	10.	4 o'clock

Lesson 12: Recognize halves within a circular clock face and tell time to the half hour.

Lesson 12 Homework 1•5

11. Match the pictures with the clocks.

a.  Soccer practice 3:30

b.  Brush teeth 7:30

c.  Wash dishes 6:00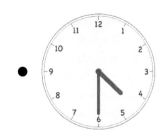

d.  Eat dinner 5:30

e.  Take bus home 4:30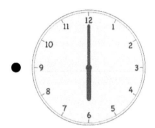

f. Homework half past 6 o'clock

Lesson 12: Recognize halves within a circular clock face and tell time to the half hour.

Lesson 13

Objective: Recognize halves within a circular clock face and tell time to the half hour.

Suggested Lesson Structure

- ■ Fluency Practice   (15 minutes)
- ▨ Application Problem   (5 minutes)
- ▢ Concept Development   (30 minutes)
- ■ Student Debrief   (10 minutes)
-    **Total Time**   **(60 minutes)**

Fluency Practice (15 minutes)

- Grade 1 Core Fluency Sprint **1.OA.6**   (10 minutes)
- Happy Counting **1.NBT.1**   (2 minutes)
- Analogous Addition and Subtraction **1.OA.6, 1.NBT.2**   (3 minutes)

Grade 1 Core Fluency Sprint (10 minutes)

Materials: (S) Core Fluency Sprint (Lesson 1 Core Fluency Sprint)

Note: Choose an appropriate Sprint, based on the needs of the class. If the majority of students completed the first three quadrants on the previous Sprint, move to the next Sprint listed in the sequence provided below (Core Fluency Sprint List). If many students are not making it to the third quadrant, consider repeating the same Sprint. As students work, pay attention to their strategies and the number of problems they are answering to consider for future Sprint administration.

Core Fluency Sprint List:

- Core Addition Sprint 1 (Targets core addition and missing addends.)
- Core Addition Sprint 2 (Targets the most challenging addition within 10.)
- Core Subtraction Sprint (Targets core subtraction.)
- Core Fluency Sprint: Totals of 5, 6, and 7 (Develops understanding of the relationship between addition and subtraction.)
- Core Fluency Sprint: Totals of 8, 9, and 10 (Develops understanding of the relationship between addition and subtraction.)

Lesson 13: Recognize halves within a circular clock face and tell time to the half hour.

A STORY OF UNITS   Lesson 13 1•5

Happy Counting (2 minutes)

Note: This activity prepares students for Module 6 by providing practice counting by ones and tens within 100.

Repeat the activity from Lesson 11.

Analogous Addition and Subtraction (3 minutes)

Note: This activity practices Grade 1's core fluency and reminds students to use their knowledge of sums and differences within 10 (e.g., 5 + 3 = 8) to solve analogous problems within 40 (e.g., 15 + 3 = 18, 25 + 3 = 28, and 35 + 3 = 38).

T: On my signal, say the equation with the answer. 6 + 2 = ____. (Pause. Signal.)
S: 6 + 2 = 8.
T: 16 + 2 = ____. (Pause. Signal.)
S: 16 + 2 = 18.

Continue with 26 + 2 and 36 + 2. Then repeat, beginning with other addition or subtraction sentences within 10.

Application Problem (5 minutes)

Ben is a clock collector. He has 8 digital clocks and 5 circular clocks. How many clocks does Ben have altogether? How many more digital clocks does Ben have than circular clocks?

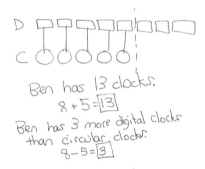

Note: Today's Application Problem is very similar to the problem in Lesson 11. Use this opportunity to recognize students who are showing improvement in solving *compare with difference unknown* problem types. Analyze students' work to pinpoint challenges and then adjust, extend, or modify Module 6 Lesson 1 to support students' development with these problem types.

Concept Development (30 minutes)

Materials: (T) Clock image 1 (Template 1) (S) Clock images (Template 2), personal white board

Note: Today's objective extends to clocks students may encounter. If the majority of the class requires more exposure to the traditional analog clock used during Lessons 10–12, substitute the variety of clock faces with the paper clock template in Lesson 11, and have students erase and redraw clock hands for each time they are given.

NOTES ON MULTIPLE MEANS OF ENGAGEMENT:

Remember to ask students, "What time is it?" throughout the day to accustom them to looking at a clock and noticing when events happen during their day. Continuing to incorporate clocks into all teaching helps students master telling time to the hour and half hour.

If using various clock faces, it might be preferable to bring in actual clocks and watches to use during the lesson, or ask families in advance to send in pictures of clocks in their homes.

Have students place the clocks template (Template 2) into their personal white boards and gather in the meeting area.

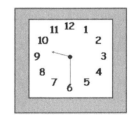

T: (Project clock image 1, a watch.) Many people use something like this to show them what time it is. Do you know what it is called?

S: A watch!

T: Why do people sometimes wear watches? Talk with your partner about it. (Wait as students share their thoughts.)

S: It tells them the time. → It's like having a clock with you even when you're outside. → People have watches because they can't carry around a big clock.

T: What is the time on this watch?

S: 3:30.

T: This watch looks a lot like the clocks we have been looking at. But sometimes watches and clocks look different from each other. What differences do you notice among the clocks and watches on the clock page in your personal white board (Template 2)?

S: One of them is a square. → Some of them have no numbers. → Some of them have a few of the numbers, but not all of the numbers. → One of them has weird letters where the numbers should be. → Some of them have pointy arrows on the clock hands.

T: Let's use what we know about circles and clocks to help us tell the time, even when the clock face looks different.

T: Let's look at the square clock. What is the time?

S: 9:30.

T: We can also say...?

S: Half past 9.

T: Write the time on the line under the clock. (Wait as students write 9:30.)

T: Let's all look at the next clock. This clock only has four numbers—3, 6, 9, and 12. Where do you think the missing numbers would go? Use your dry erase marker to put them in. (Wait as students place in the missing numbers.)

T: What time does the clock show?

S: 4 o'clock!

NOTES ON MULTIPLE MEANS OF ENGAGEMENT:

While teaching, be sure to provide cross-curricular connections for students. Interested students can write a story about their school day and the events that occur at certain times. During other writing activities, encourage students to incorporate time into their stories.

T: Write the time on the line under the clock.

T: Try the next clock without putting in the missing numbers. *Imagine* the numbers that are missing. (Wait as students write the time on the line.)

T: What time does the clock show?

S: 11 o'clock!

T: How did you know?

S: The minute hand was on the 12, and the hour hand was just before that, so it had to be the hour that's before 12, which is 11 o'clock.

Repeat the process for each watch or clock on the page, discussing ways to determine the time based on the position of the hands on the clock face.

Problem Set (10 minutes)

Students should do their personal best to complete the Problem Set within the allotted 10 minutes. For some classes, it may be appropriate to modify the assignment by specifying which problems they work on first.

Student Debrief (10 minutes)

Lesson Objective: Recognize halves within a circular clock face and tell time to the half hour.

The Student Debrief is intended to invite reflection and active processing of the total lesson experience.

Invite students to review their solutions for the Problem Set. They should check work by comparing answers with a partner before going over answers as a class. Look for misconceptions or misunderstandings that can be addressed in the Debrief. Guide students in a conversation to debrief the Problem Set and process the lesson.

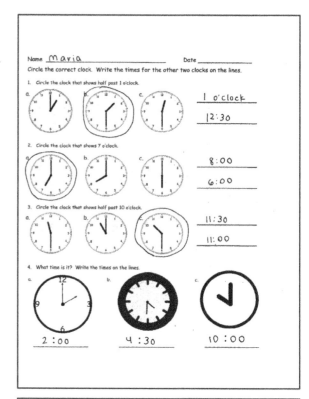

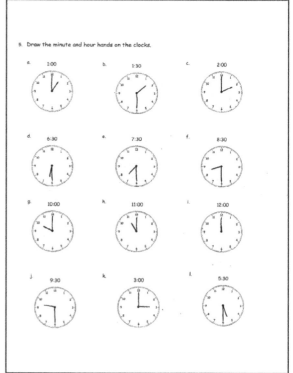

Any combination of the questions below may be used to lead the discussion.

- Look at your Problem Set. Which clock was the most challenging for you to read and why?
- Look at the clocks on your personal white board. Which clock was the most challenging for you to read and why? Which clock would you like to have in your home and why?
- No matter what a clock looks like, what parts must it include in order for us to tell the time?
- When can it be helpful to know what time it is?
- Look at the Application Problem. Share how you used your drawing to help solve the problem.

Exit Ticket (3 minutes)

After the Student Debrief, instruct students to complete the Exit Ticket. A review of their work will help with assessing students' understanding of the concepts that were presented in today's lesson and planning more effectively for future lessons. The questions may be read aloud to the students.

A STORY OF UNITS

Lesson 13 Problem Set 1•5

Name _____ Date _____

Circle the correct clock. Write the times for the other two clocks on the lines.

1. Circle the clock that shows half past 1 o'clock.

a. 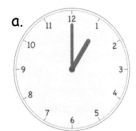   b.    c.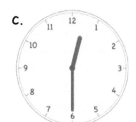

_____

_____

2. Circle the clock that shows 7 o'clock.

a. 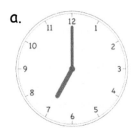   b.    c.

_____

_____

3. Circle the clock that shows half past 10 o'clock.

a. 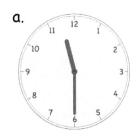   b. 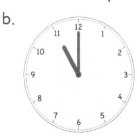   c.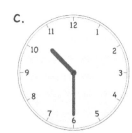

_____

_____

4. What time is it? Write the times on the lines.

a. 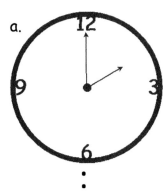   b. 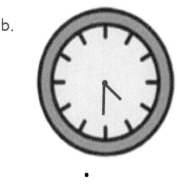   c.

____:____   ____:____   ____:____

Lesson 13: Recognize halves within a circular clock face and tell time to the half hour.

5. Draw the minute and hour hands on the clocks.

a. 1:00

b. 1:30

c. 2:00

d. 6:30

e. 7:30

f. 8:30

g. 10:00

h. 11:00

i. 12:00

j. 9:30

k. 3:00

l. 5:30

Name _____   Date _____

1. Circle the clock(s) that shows half past 3 o'clock.

   a.    b.   c.

2. Write the time or draw the hands on the clocks.

   a.    b.   c.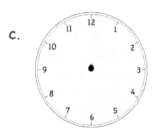

   4:30   9 o'clock
                              _____

Name _____   Date _____

Fill in the blanks.

1.  Clock _____ shows half past three.

2.   Clock _____ shows half past twelve.

3.   Clock _____ shows eleven o'clock.

4.   Clock _____ shows 8:30.

5.   Clock _____ shows 5:00.

Lesson 13 Homework 1•5

6. Write the time on the line under the clock.

a. 12:00	b. 6:00	c. 6:00
d. 7:30	e. 6:00	f. 2:00
g. 3:30	h. 11:00	i. 2:30

7. Put a check (✓) next to the clock(s) that show 4 o'clock.

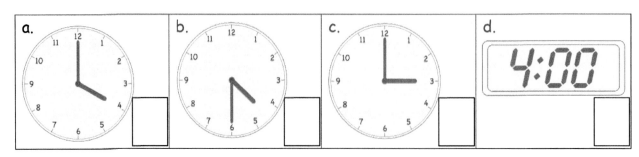

a. b. c. ✓ d. ✓

clock image 1

clock images

A STORY OF UNITS End-of-Module Assessment Task 1•5

Name _____ Date _____

1. Color the shapes using the key. Write how many of each shape there are on the line.

a. YELLOW Circles: _____

b. RED   Rectangles: _____

c. BLUE   Triangles: _____

d. GREEN   Trapezoids: _____

e. BLACK   Hexagons: _____

f. ORANGE   Rhombuses: _____

2. Is the shape a triangle?
   If it is, write YES on the line. If it is not, explain why it is not a triangle on the line.

a.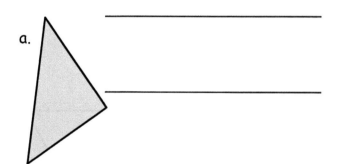

b.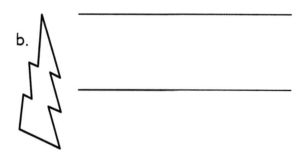

c.

d.

Module 5: Identifying, Composing, and Partitioning Shapes

3. a. Circle the attributes that are used to describe *all* cylinders.

Cylinders can roll.	Cylinders are hollow.
Cylinders are made of paper.	Cylinders have 2 flat surfaces made of circles or ovals.

b. Circle the attributes that are used to describe *all* rectangular prisms.

Rectangular prisms can roll.	The faces of a rectangular prism are rectangles.
Rectangular prisms have 6 faces.	Rectangular prisms are made of wood.

4. Use your triangle pattern blocks to cover the shapes below. Draw lines to show how you formed the shape with your triangles.

a.

b.

A STORY OF UNITS

End-of-Module Assessment Task 1•5

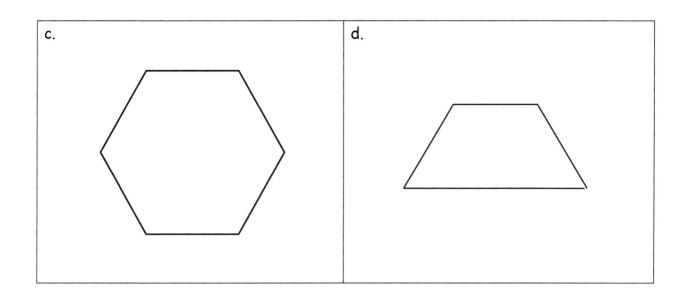

e. Here are the pieces that Dana is putting together to create a shape.

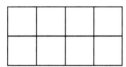   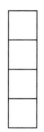

Which of the following shows what Dana's shape might look like when she combines her smaller shapes?

5.
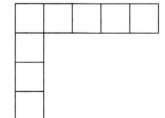   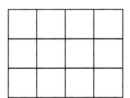   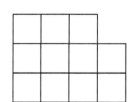

Module 5: Identifying, Composing, and Partitioning Shapes

5. Match the time to the correct clock.

   a. ten o'clock   `1:00`

   b. ten thirty   `10:00`

   c. one o'clock   `10:30`

   d. three thirty   `3:30`

6. Write the time on the line.

a. _____   b. _____   c. _____

   d. Circle the clock that shows half past 5 o'clock.

7. Draw the minute hand so that the clock shows the time written above it.

   a.   4:30

   b.   5:00

   c. Draw one line to make this rectangle into two squares that are the same size.

   d. Circle the words that make the sentence true.

   One square makes up (**one half** / **one quarter**) of the rectangle above.

   e. Color one half of the rectangle. What shapes were used to make the rectangle?

   f. Color one fourth of the rectangle. What shapes were used to make the rectangle?

g. Color one fourth of the circle. The dot is in the center.

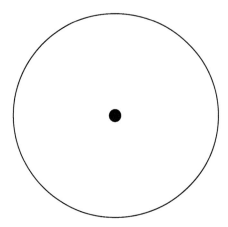

A STORY OF UNITS

End-of-Module Assessment Task 1•5

| End-of-Module Assessment Task Standards Addressed | Topics A–D |

Tell and write time and money.[1]

1.MD.3   Tell and write time in hours and half-hours using analog and digital clocks. Recognize and identify coins, their names, and their values.

Reason with shapes and their attributes.

1.G.1   Distinguish between defining attributes (e.g., triangles are closed and three-sided) versus non-defining attributes (e.g., color, orientation, overall size); build and draw shapes to possess defining attributes.

1.G.2   Compose two-dimensional shapes (rectangles, squares, trapezoids, triangles, half-circles, and quarter-circles) or three-dimensional shapes (cubes, right rectangular prisms, right circular cones, and right circular cylinders) to create a composite shape, and compose new shapes from the composite shape. (Students do not need to learn formal names such as "right rectangular prism.")

1.G.3   Partition circles and rectangles into two and four equal shares, describe the shares using the words *halves, fourths*, and *quarters*, and use the phrases *half of, fourth of*, and *quarter of*. Describe the whole as two of, or four of the shares. Understand for these examples that decomposing into more equal shares creates smaller shares.

Evaluating Student Learning Outcomes

A Progression Toward Mastery is provided to describe steps that illuminate the gradually increasing understandings that students develop *on their way to proficiency.* In this chart, this progress is presented from left (Step 1) to right (Step 4). The learning goal for students is to achieve Step 4 mastery. These steps are meant to help teachers and students identify and celebrate what the students CAN do now and what they need to work on next.

---

[1]Time alone is addressed in this module. Money is addressed in Module 6.

Module 5:   Identifying, Composing, and Partitioning Shapes

End-of-Module Assessment Task 1•5

A Progression Toward Mastery				
Assessment Task Item and Standards Assessed	STEP 1 Little evidence of reasoning without a correct answer. (1 Point)	STEP 2 Evidence of some reasoning without a correct answer. (2 Points)	STEP 3 Evidence of some reasoning with a correct answer or evidence of solid reasoning with an incorrect answer. (3 Points)	STEP 4 Evidence of solid reasoning with a correct answer. (4 Points)
1 1.G.1	The student identifies the correct number for fewer than four of the six shapes.	The student identifies the correct number for at least four of the six shapes.	The student identifies the correct number for five of the six shapes.	The student correctly colors and provides the following counts: - Circles: 2 - Rectangles: 3 - Triangles: 2 - Trapezoids: 2 - Hexagons: 1 - Rhombuses: 1* (*Some students may include square as well.)
2 1.G.1	The student does not identify (a) and (d) as triangles.	The student correctly identifies (a) and (d) as triangles but does not clearly explain why both of the other two shapes are not triangles.	The student correctly identifies (a) and (d) as triangles but only explains why one of the other two shapes is not a triangle.	The student correctly writes: a. Yes. b. It has more than three sides. c. It is not closed. Or, it has less than three sides. d. Yes.
3 1.G.1	The student circles three or more incorrect sentences, or the student circles fewer than two correct answers.	The student correctly circles at least two of the four correct answers and may circle one or two incorrect sentences.	The student circles at least three of the four correct answers and only circles one incorrect sentence.	The student correctly circles the following choices: a. Cylinders can roll. Cylinders have two flat surfaces made of circles or ovals. b. Rectangular prisms have 6 faces. The faces of a rectangular prism are rectangles.

A Story of Units
End-of-Module Assessment Task 1•5

A Progression Toward Mastery					
4 1.G.2	The student is unable to demonstrate understanding of accurately composing the given shape using triangles and did not choose the middle image for part (e).	The student correctly draws partitions for one or two parts. The student may have chosen the middle image for part (e).	The student is able to correctly draw partitions for at least three parts and chooses the middle image for part (e).	The student draws lines to show: a. 2 triangles b. 4 triangles c. 6 triangles d. 3 triangles The student chooses the middle image.
5 1.MD.3	The student is unable to demonstrate understanding of telling time from a digital clock and is unable to match any of the times.	The student demonstrates limited understanding of telling time from a digital clock, matching one time correctly.	The student demonstrates understanding of telling either the hour or the minutes from a digital clock, matching two or three times correctly.	The student correctly matches: a. 10:00 b. 10:30 c. 1:00 d. 3:30
6 1.MD.3	The student is unable to demonstrate understanding of telling time from an analog clock, answering none or one part correctly.	The student demonstrates understanding of telling time to the hour from an analog clock, answering one or two parts correctly.	The student demonstrates understanding of telling time to the hour from an analog clock and some ability to tell time to the half hour, answering three parts correctly OR correctly stating the numerals for all times but missing *o'clock* in (a) and (b).	The student correctly writes: a. One o'clock b. Six o'clock c. One thirty d. Choice 3 (Spelling is not being assessed. Students may write the time using digital notation, as shown in the sample, or as written above.)

Module 5: Identifying, Composing, and Partitioning Shapes

A Story of Units — End-of-Module Assessment Task 1•5

A Progression Toward Mastery

7 1.MD.3 1.G.1 1.G.2 1.G.3	The student answers none to two of the seven parts correctly.	The student is able to complete at least three of the seven parts correctly.	The student is able to complete at least five of the seven parts correctly, OR has up to three slight errors in approximating halves or fourths when coloring.	The student correctly: a. Draws a minute hand pointing to 6. b. Draws a minute hand pointing to 12. c. Draws a line to create two squares. d. Circles *one half*. e. Colors a triangle and writes *triangle*. f. Colors one rectangle and writes *rectangle* (or *rectangles and squares*). g. Colors one fourth of the circle.

Module 5: Identifying, Composing, and Partitioning Shapes

A STORY OF UNITS   End-of-Module Assessment Task   1•5

Name Maria _____   Date _____

1. Color the shapes using the key. Write how many of each shape there are on the line.

a. YELLOW   Circles: 2
b. RED   Rectangles: 3
c. BLUE   Triangles: 2
d. GREEN   Trapezoids: 2
e. BLACK   Hexagons: 1
f. ORANGE   Rhombuses: 1

2. Is the shape a triangle?
   If it is, write YES on the line. If it is not, explain why it is not a triangle on the line.

a. 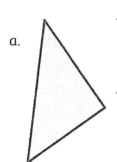 Yes _____

b.  No, it has more than 3 sides.

c. 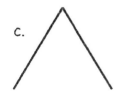 No, it is not closed. It has 2 sides.

d. 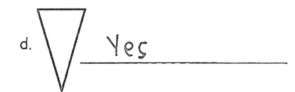 Yes _____

Module 5:   Identifying, Composing, and Partitioning Shapes

3. a. Circle the attributes that are used to describe **all** cylinders.

(Cylinders can roll.)	Cylinders are hollow.
Cylinders are made of paper.	(Cylinders have 2 flat surfaces made of circles or ovals.)

b. Circle the attributes that are used to describe **all** rectangular prisms.

Rectangular prisms can roll.	(The faces of a rectangular prism are rectangles.)
(Rectangular prisms have 6 faces.)	Rectangular prisms are made of wood.

4. Use your triangle pattern blocks to cover the shapes below. Draw lines to show how you formed the shape with your triangles.

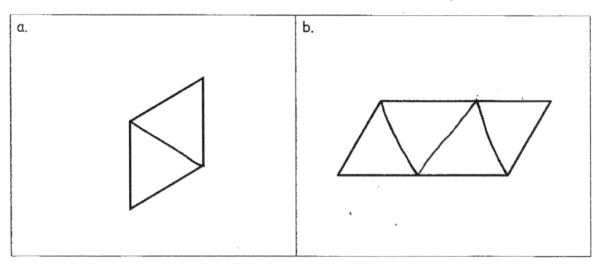

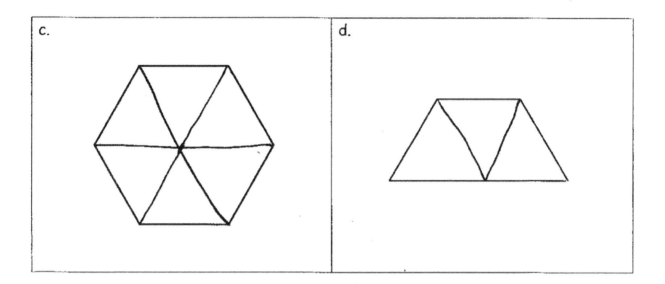

e. Here are the pieces that Dana is putting together to create a shape.

   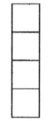

Which of the following shows what Dana's shape might look like when she combines her smaller shapes?

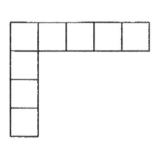   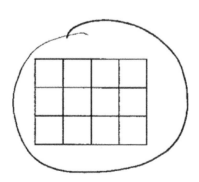   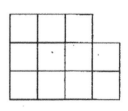

5. Match the time to the correct clock.

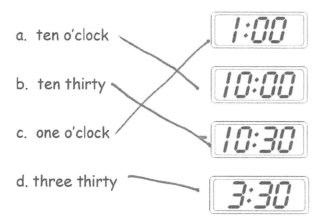

a. ten o'clock
b. ten thirty
c. one o'clock
d. three thirty

1:00
10:00
10:30
3:30

6. Write the time on the line.

a. 1:00   b. 6:00   c. 1:30

d. Circle the clock that shows half past 5 o'clock.

7. Draw the minute hand so that the clock shows the time written above it.

a. 4:30

b. 5:00

c. Draw one line to make this rectangle into two squares that are the same size.

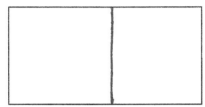

d. Circle the words that make the sentence true.

One square makes up ((one half)/ one quarter) of the rectangle above.

e. Color one half of the rectangle. What shapes were used to make the rectangle?

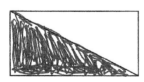   triangle

f. Color one fourth of the rectangle. What shapes were used to make the rectangle?

   rectangle

g. Color one fourth of the circle. The dot is in the center.

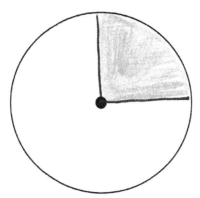

Answer Key

Eureka Math
Grade 1
Module 5

Special thanks go to the Gordon A. Cain Center and to the Department of Mathematics at Louisiana State University for their support in the development of *Eureka Math*.

For a free *Eureka Math* Teacher Resource Pack, Parent Tip Sheets, and more please visit www.Eureka.tools

Published by the non-profit Great Minds

Copyright © 2015 Great Minds. No part of this work may be reproduced, sold, or commercialized, in whole or in part, without written permission from Great Minds. Non-commercial use is licensed pursuant to a Creative Commons Attribution-NonCommercial-ShareAlike 4.0 license; for more information, go to http://greatminds.net/maps/math/copyright. "Great Minds" and "Eureka Math" are registered trademarks of Great Minds.

Printed in the U.S.A.

This book may be purchased from the publisher at eureka-math.org

10 9 8 7 6 5 4

ISBN 978-1-63255-352-2

A STORY OF UNITS

Mathematics Curriculum

GRADE

GRADE 1 • MODULE 5

Answer Key
GRADE 1 • MODULE 5
Identifying, Composing, and Partitioning Shapes

Lesson 1

Core Addition Sprint 1

Side A

1. 5
2. 6
3. 7
4. 7
5. 8
6. 9
7. 6
8. 7
9. 8
10. 3
11. 5
12. 9
13. 10
14. 3
15. 3
16. 7
17. 3
18. 3
19. 9
20. 4
21. 5
22. 9
23. 7
24. 2
25. 9
26. 6
27. 3
28. 6
29. 6
30. 2

Side B

1. 6
2. 7
3. 8
4. 5
5. 6
6. 7
7. 4
8. 5
9. 6
10. 3
11. 3
12. 7
13. 8
14. 3
15. 5
16. 6
17. 2
18. 2
19. 7
20. 4
21. 3
22. 9
23. 5
24. 4
25. 8
26. 3
27. 7
28. 2
29. 2
30. 3

Core Addition Sprint 2

Side A

1. 7
2. 8
3. 9
4. 7
5. 8
6. 9
7. 8
8. 8
9. 9
10. 9
11. 7
12. 7
13. 8
14. 8
15. 7
16. 9
17. 9
18. 9
19. 9
20. 8
21. 9
22. 7
23. 9
24. 9
25. 7
26. 17
27. 17
28. 9
29. 6
30. 3

Side B

1. 7
2. 8
3. 9
4. 8
5. 9
6. 9
7. 7
8. 7
9. 8
10. 8
11. 8
12. 8
13. 9
14. 9
15. 9
16. 9
17. 9
18. 9
19. 9
20. 9
21. 8
22. 7
23. 9
24. 9
25. 9
26. 19
27. 19
28. 9
29. 5
30. 4

Core Subtraction Sprint

Side A

1. 5
2. 4
3. 3
4. 9
5. 8
6. 7
7. 5
8. 6
9. 7
10. 4
11. 5
12. 7
13. 6
14. 5
15. 4
16. 6
17. 2
18. 4
19. 3
20. 4
21. 5
22. 4
23. 3
24. 3
25. 7
26. 6
27. 7
28. 6
29. 7
30. 8

Side B

1. 4
2. 3
3. 2
4. 9
5. 8
6. 7
7. 4
8. 5
9. 6
10. 3
11. 4
12. 5
13. 1
14. 2
15. 3
16. 4
17. 2
18. 5
19. 3
20. 2
21. 6
22. 3
23. 2
24. 4
25. 5
26. 4
27. 5
28. 7
29. 9
30. 7

Core Fluency Sprint: Totals of 5, 6, & 7

Side A

1. 5
2. 2
3. 2
4. 3
5. 3
6. 6
7. 5
8. 5
9. 1
10. 1
11. 6
12. 4
13. 4
14. 2
15. 2
16. 6
17. 3
18. 3
19. 7
20. 2
21. 5
22. 2
23. 2
24. 7
25. 3
26. 3
27. 4
28. 4
29. 6
30. 6

Side B

1. 5
2. 1
3. 1
4. 4
5. 4
6. 7
7. 2
8. 5
9. 2
10. 5
11. 6
12. 5
13. 5
14. 1
15. 1
16. 6
17. 3
18. 3
19. 6
20. 2
21. 4
22. 2
23. 2
24. 7
25. 3
26. 3
27. 3
28. 3
29. 7
30. 6

Lesson 1 Answer Key

Core Fluency Sprint: Totals of 8, 9, & 10

Side A

1. 10
2. 5
3. 5
4. 10
5. 9
6. 9
7. 1
8. 1
9. 9
10. 1
11. 8
12. 1
13. 8
14. 8
15. 4
16. 8
17. 2
18. 6
19. 9
20. 2
21. 2
22. 6
23. 2
24. 9
25. 3
26. 3
27. 6
28. 6
29. 10
30. 9

Side B

1. 10
2. 9
3. 9
4. 1
5. 1
6. 8
7. 1
8. 7
9. 1
10. 7
11. 10
12. 8
13. 8
14. 2
15. 2
16. 8
17. 3
18. 5
19. 8
20. 2
21. 2
22. 9
23. 7
24. 2
25. 9
26. 4
27. 4
28. 4
29. 8
30. 8

A STORY OF UNITS   Lesson 1 Answer Key 1•5

Problem Set

1. 3 pentagons circled
2. 1 circle and 3 ovals circled
3. 1 rectangle and 1 square circled
4. a. Triangle drawn
   b. Different triangle drawn

5. Have 3 straight sides; have 3 corners
6. Triangle circled
7. Two triangles drawn
8. Any shape without 3 sides and 3 corners drawn

Exit Ticket

1. a. 3; 3
   b. 4; 4
   c. 0; 0

2. a. Square crossed off
   b. Parallelogram crossed off

Homework

1. 4 triangles circled
2. 1 circle and 3 ovals circled
3. 1 rectangle and 1 square circled
4. a. Answers will vary.
   b. Answers will vary.

5. Have 4 straight sides; have 4 square corners
6. Rectangle circled
7. Two rectangles drawn
8. Any shape that is not a rectangle drawn

Module 5:   Identifying, Composing, and Partitioning Shapes

Lesson 2

Problem Set

1. a. 7
   b. 3
   c. 0
   d. 1
   e. 7

2. 3 rectangles circled

3. a. Yes, explanations will vary.
   b. No, explanations will vary.

Exit Ticket

1. 0; 0; circle
2. 3; 3; triangle
3. 6; 6; hexagon
4. 4; 4; rectangle, square, rhombus

Homework

1. 3; 4; 1; 2
2. a. 3; 3
   b. 3
3. a. 6; 6
   b. 1
4. a. 0; 0
   b. 2
5. a. 4; 4
   b. 1
6. a. Fourth shape from left crossed off
   b. Open shape; it only has 3 corners
7. a. Fourth shape from left crossed off
   b. All four sides are not of equal length

Lesson 3

Core Fluency Differentiated Practice Set A

1. 6
2. 6
3. 6
4. 6
5. 7
6. 7
7. 8
8. 7
9. 7
10. 6
11. 8
12. 8
13. 6
14. 7
15. 8
16. 9
17. 10
18. 9
19. 9
20. 10
21. 8
22. 9
23. 10
24. 10
25. 8
26. 7
27. 10
28. 9
29. 10
30. 9

Core Fluency Differentiated Practice Set B

1. 0
2. 6
3. 1
4. 2
5. 7
6. 1
7. 6
8. 1
9. 7
10. 2
11. 3
12. 4
13. 5
14. 4
15. 2
16. 3
17. 4
18. 2
19. 3
20. 4
21. 3
22. 4
23. 5
24. 6
25. 7
26. 8
27. 7
28. 6
29. 5
30. 6

Core Fluency Differentiated Practice Set C

1. 1
2. 5
3. 5
4. 1
5. 9
6. 1
7. 5
8. 5
9. 2
10. 2
11. 3
12. 3
13. 2
14. 2
15. 3
16. 3
17. 2
18. 2
19. 3
20. 3
21. 4
22. 4
23. 3
24. 3
25. 4
26. 4
27. 3
28. 3
29. 3
30. 3

Core Fluency Differentiated Practice Set D

1. 6
2. 5
3. 6
4. 7
5. 4
6. 5
7. 7
8. 0
9. 1
10. 3
11. 3
12. 4
13. 6
14. 2
15. 4
16. 6
17. 5
18. 1
19. 2
20. 2
21. 4
22. 5
23. 3
24. 4
25. 5
26. 4
27. 3
28. 2
29. 3
30. 2

Lesson 3 Answer Key

Core Fluency Differentiated Practice Set E

1. 6
2. 4
3. 3
4. 7
5. 2
6. 7
7. 3
8. 4
9. 9
10. 5
11. 4
12. 4
13. 2
14. 2
15. 2
16. 3
17. 4
18. 2
19. 6
20. 2
21. 3
22. 5
23. 3
24. 7
25. 2
26. 7
27. 2
28. 3
29. 5
30. 3

Problem Set

1.
   a. Cone
   b. Cube
   c. Cylinder
   d. Rectangular prism
   e. Sphere

2. Cubes: block, dice

   Spheres: globe, tennis ball

   Cones: party hat

   Rectangular Prisms: tissue box

   Cylinders: can

3. Have no straight sides; are round; can roll circled

4. Have square faces; have 6 faces circled

Exit Ticket

1. True. Answers may vary.
2. False. Answers may vary.

Homework

1. Answers may vary.
2. Answers may vary.

Lesson 4

Problem Set

1. Drawing of trapezoid made with 3 triangles
2. Drawing of large square made with 4 squares
3. Drawing of hexagon made with 6 triangles
4. Drawing of hexagon made with 1 trapezoid, 1 rhombus, 1 triangle
5. Answers may vary.
6. 50
7. Answers may vary.

Exit Ticket

1. Drawing of hexagon made with 3 rhombuses
2. Drawing of triangle made with 1 hexagon and 3 triangles

Homework

1. Drawing of trapezoid made with 3 blue triangles
2. Drawing of hexagon made with 3 blue triangles and 1 green trapezoid
3. 30

Lesson 5

Problem Set

1. a. 7
   b. triangle, square, parallelogram
2. Square made with 2 triangles drawn
3. Trapezoid made with 4 tangram pieces drawn
4. Puzzle completed
5. Answers will vary.

Exit Ticket

Answers will vary.

Homework

1. Answers will vary.
2. Answers will vary.
3. a. Triangle made with 2 small triangles drawn
   b. Trapezoid made with 1 square and 1 triangle drawn
   c. Answers will vary.
4. Answers will vary.

Lesson 6

Problem Set

1. Answers will vary.
2. Answers will vary.
3. Answers will vary.
4. Answers will vary.

Exit Ticket

Structure matching Maria's structure built

Homework

Answers will vary.

Lesson 7

Problem Set

1. a. Answer provided
    b. N
    c. N
    d. Y, 4
    e. Y, 2
    f. N
    g. Y, 4
    h. Y, 2
    i. N
    j. Y, 2
    k. Y, 2
    l. Y, 6
    m. Y, 2
    n. N
    o. Y, 2

2. a. 2
    b. 4
    c. 2
    d. 2
    e. 2
    f. 4

3. Vertical line drawn from corner to midpoint

4. Answers may vary.

5. Lines drawn accurately

Exit Ticket

Circle; 4

Homework

1. a. Answer provided
    b. Y, 2
    c. N
    d. N
    e. Y, 2
    f. Y, 4
    g. Y, 2
    h. N
    i. N
    j. Y, 2
    k. N
    l. Y, 4
    m. Y, 6
    n. Y, 2
    o. N

2. Line drawn accurately; triangles

3. Line drawn accurately; rectangles

4. Lines drawn accurately; triangles

Lesson 8

Problem Set

1. a. No
   b. No
   c. Yes
   d. Yes
   e. No
   f. Yes
2. a. Yes
   b. Yes
   c. No
   d. Yes
   e. No
   f. No
3. a. Answers may vary.
   b. Answers may vary.
   c. Answers may vary.
   d. Answers may vary.
   e. Answers may vary.
   f. Answers may vary.
4. a. Answers may vary.
   b. Answers may vary.
   c. Answers may vary.
   d. Answers may vary.
   e. Answers may vary.
   f. Answers may vary.

Exit Ticket

Answers may vary.

Homework

1. a. Unequal parts
   b. Equal parts
   c. Halves
   d. Quarters
   e. Quarters
   f. Halves
   g. Quarters
   h. Fourths
2. a. 1 half
   b. 1 quarter
   c. 1 half
   d. 1 half
3. Answers may vary.
4. Answers may vary.

Lesson 9

Problem Set

1. A; B; A
2. B; A; B
3. Second rectangle circled; one half of circled
4. Is larger than circled
5. Is smaller than circled
6. Is the same size as circled

Exit Ticket

1. a. F
   b. T
2. Answers may vary.

Homework

1. A; B; A
2. a. Half; half
   b. Quarter; half
   c. Quarter; quarter
   d. Half; half
3. a. Is smaller than circled
   b. Is the same size as circled

Lesson 10

Problem Set

1. a. 5:00
   b. 12:00
   c. 8:00
   d. 1:00

2. Hour hand drawn pointing at 3

3. a. 5:00
   b. 1
   c. 3
   d. 9
   e. 12:00
   f. 7
   g. 4:00
   h. 6
   i. 11:00
   j. 10
   k. 6:00
   l. 2
   m. 11 o'clock
   n. 8:00 or 8 o'clock
   o. 3:00 or 3 o'clock

Exit Ticket

1. 6:00 or 6 o'clock
2. 9:00 or 9 o'clock
3. 7:00 or 7 o'clock
4. 12:00 or 12 o'clock

Homework

1. a. 2 o'clock, 2:00
   b. 7 o'clock, 7:00
   c. 11 o'clock, 11:00
   d. 4 o'clock, 4:00
   e. 10 o'clock, 10:00
   f. 3 o'clock, 3:00

2. a. Answer provided
   b. Hand pointing to 9, 9:00
   c. Hand pointing to 12, 12:00
   d. Hand pointing to 7, 7:00
   e. Hand pointing to 1, 1:00

# Lesson 11

Problem Set

1. a. 2:30, Two thirty
   b. Half past 5 o'clock, Five thirty
   c. 12:30, half past 12 o'clock
2. a. Minute hand drawn pointing to 12
   b. Minute hand drawn pointing to 12
   c. Minute hand drawn pointing to 6
   d. Minute hand drawn pointing to 6
   e. Minute hand drawn pointing to 6
   f. Minute hand drawn pointing to 12
3. a. Answer provided
   b. Answer provided
   c. 11:30, eleven thirty, or half past 11 o'clock
   d. 2:30, two thirty, or half past 2 o'clock
   e. 2:00 or 2 o'clock
   f. 8:30, eight thirty, or half past 8 o'clock
   g. 10:30, ten thirty, or half past 10 o'clock
   h. 6:30, six thirty, or half past 6 o'clock
   i. 7:00 or 7 o'clock
   j. Seven thirty or half past 7 o'clock
   k. 4:30, four thirty, or half past 4 o'clock
   l. Ten thirty or half past 10 o'clock
4. Clock (c.)

Exit Ticket

1. Minute hand drawn pointing to 6
2. Minute hand drawn pointing to 6
3. 1:30, one thirty, or half past 1 o'clock

Homework

1. Clock (b.)
2. Clock (a.)
3. Clock (a.)
4. Clock (b.)
5. 6:30, six thirty, or half past 6 o'clock
6. 7:30, seven thirty, or half past 7 o'clock
7. 10:30, ten thirty, or half past 10 o'clock
8. 12:30, twelve thirty, or half past 12 o'clock
9. 3:30, three thirty, or half past 3 o'clock
10. 4:30, four thirty, or half past 4 o'clock
11. 5:30, five thirty, or half past 5 o'clock
12. 7:30, seven thirty, or half past 7 o'clock

A STORY OF UNITS   Lesson 12 Answer Key   1•5

Lesson 12

Problem Set

1. A
2. A
3. A
4. B
5. B
6. a. 7 o'clock, 7:00
   b. Half past 5, 5:30
   c. Half past 1, 1:30
   d. Half past 7, 7:30

7. a. Hour hand between 3 and 4, minute hand at 6
   b. Hour hand between 8 and 9, minute hand at 6
   c. Hour hand at 11, minute hand at 12
   d. Hour hand at 6, minute hand at 12
   e. Hour hand between 4 and 5, minute hand at 6
   f. Hour hand between 12 and 1, minute hand at 6

Exit Ticket

1. Hour hand between 1 and 2, minute hand at 6
2. Hour hand at 10, minute hand at 12
3. Hour hand between 5 and 6, minute hand at 6
4. Hour hand between 7 and 8, minute hand at 6

Homework

1. Hour hand at 10, minute hand at 12
2. Minute hand at 6
3. Hour hand at 8
4. 6:30 or half past 6 o'clock
5. Hour hand at 3, minute hand at 12
6. Minute hand at 6
7. 2:00 or two o'clock
8. Hour hand between 6 and 7, minute hand at 6

9. Minute hand at 6
10. Hour hand at 4, minute hand at 12
11. a. 3:30—second clock
    b. 7:30—sixth clock
    c. 6:00—fifth clock
    d. 5:30—first clock
    e. 4:30—fourth clock
    f. Half past 6 o'clock—third clock

Module 5:   Identifying, Composing, and Partitioning Shapes   219

Lesson 13

Problem Set

1. Clock (b.) circled; 1:00 or 1 o'clock; 12:30, twelve thirty, or half past twelve o'clock
2. Clock (a.) circled; 8:00 or 8 o'clock; 6:00 or six o'clock
3. Clock (c.) circled; 11:30, eleven thirty, or half past 11 o'clock; 11:00 or 11 o'clock
4. a. 2:00
   b. 4:30
   c. 10:00
5. a. Hour hand at 1, minute hand at 12
   b. Hour hand between 1 and 2, minute hand at 6
   c. Hour hand at 2, minute hand at 12
   d. Hour hand between 6 and 7, minute hand at 6
   e. Hour hand between 7 and 8, minute hand at 6
   f. Hour hand between 8 and 9, minute hand at 6
   g. Hour hand at 10, minute hand at 12
   h. Hour hand at 11, minute hand at 12
   i. Hour hand at 12, minute hand at 12
   j. Hour hand between 9 and 10, minute hand at 6
   k. Hour hand at 3, minute hand at 12
   l. Hour hand between 5 and 6, minute hand at 6

Exit Ticket

1. Clock (c.)
2. a. Hour hand between 4 and 5, minute hand at 6
   b. 12:00 or twelve o'clock
   c. Hour hand at 9, minute hand at 12

Lesson 13 Answer Key 1•5

Homework

1. B
2. B
3. A
4. A
5. B

6.
   a. 1:00 or one o'clock
   b. 11:30, eleven thirty, or half past 11 o'clock
   c. 6:00 or six o'clock
   d. Seven thirty or half past 7 o'clock
   e. 5:30 or half past 5 o'clock
   f. 2:30, two thirty, or half past 2 o'clock
   g. 7:00 or seven o'clock
   h. Eleven o'clock
   i. 9:30, nine thirty, or half past 9 o'clock

7. Clocks (a.) and (d.)